气象科技英语教程

(第三版)

THE COURSE OF ENGLISH FOR
METEOROLOGICAL SCIENCE AND TECHNOLOGY

(Third Edition)

寿绍文　彭　广　姚永红　寿亦萱　沈新勇　编著

China Meteorological Press

内容简介

本书主要为气象专业的学生在科技英语读、写、译、听、说五个方面能力训练提供必要的材料和知识。全书共分三大部分：第一部分为气象科技英语读物，主要训练阅读、翻译和写作能力；第二部分为气象科技英语听说材料，配有相应的免费听力音频（于气象出版社官网下载，网址为：http://www.qxcbs.com/ebook/qxkjyy/mdata.html），主要训练听说能力；第三部分为科技英语知识，目的是使学生对科技英语特点有较系统的认识。

本书可作为高等院校大气科学及相关专业学生的教材，也可作为气象、海洋、航空、农林、水利、环境等部门的科研和业务人员的参考用书。

图书在版编目（CIP）数据

气象科技英语教程 / 寿绍文等编著. — 3版. — 北京：气象出版社，2019.4（2021.8重印）

ISBN 978-7-5029-6961-5

Ⅰ.①气… Ⅱ.①寿… Ⅲ.①气象学-英语-高等学校-教材 Ⅳ.①P4

中国版本图书馆CIP数据核字（2019）第079715号

出版发行：	气象出版社		
地　　址：	北京市海淀区中关村南大街46号	邮政编码：	100081
电　　话：	010-68407112（总编室）　010-68408042（发行部）		
网　　址：	http://www.qxcbs.com	E-mail：	qxcbs@cma.gov.cn
责任编辑：	黄红丽	终　　审：	吴晓鹏
责任校对：	王丽梅	责任技编：	赵相宁
封面设计：	博雅思企划		
印　　刷：	三河市百盛印装有限公司		
开　　本：	720 mm×960 mm　1/16	印　　张：	19
字　　数：	383千字		
版　　次：	2019年4月第3版	印　　次：	2021年8月第2次印刷
定　　价：	55.00元		

本书如存在文字不清、漏印以及缺页、倒页、脱页等，请与本社发行部联系调换。

第三版前言

《气象科技英语教程》(第二版)一书自2008年出版以来,至今已过去整整十年。多年来本书一直受到广大读者的好评和欢迎。最近又荣幸地被列入南京信息工程大学"三百工程"建设精品教材建设项目,同时也得到中国气象局和四川省气象局等的大力支持,成为局校共建的重点教材建设项目,因而又一次获得了宝贵的修订和再版机会。

科技英语是描述科技用语中各种语言现象和特性的一种英语体系,它是用来进行国际科技交流的重要手段。近年来由于国际气象科技交流日益频繁,气象科技英语教学越来越受到重视。本教程为气象和相关专业学生在气象科技英语读、写、译、听、说等方面的训练提供了必要材料和知识。这次修订,我们对本书第二版的原有内容作了适当增删和修订。在部分阅读材料的课文后增加了一些补充片段、短文、插图以及少量的练习和作业等内容。此外,为方便读者使用,由加拿大专家玛格丽特•帕特森(Margaret Paterson)女士和苏珊娜•莫罗(Suzanne Moreau)女士及乔治•莫罗(Georges Moreau)先生对第二部分气象科技英语听说材料的朗读内容不再以附带光盘形式提供,读者可方便地通过气象出版社官网下载相关音频文件(免费,网址为:http://www.qxcbs.com/ebook/qxkjyy/mdata.html)。参与本次修订的除了寿绍文(主编)外,还有彭广、姚永红、寿亦萱、沈新勇等。全体编著者都认真审阅了相关课文和内容,为修订做出了很多重要贡献,特别是彭广为本次修订做了很多重要工作并提供了很多创新思路,使本书增添了新的色彩。此外,还有苗峻峰、王黎娟、高庆九等教授及本课程其他的任课老师和专家们(恕不一一列名)也提供了许多宝贵的意见和建议。这里我们要向所有在本书中被引用的文章的作者们和所有给予我们大力帮助的同事、朋友以及气象出版社编辑们表示衷心的感谢。并诚挚地欢迎广大读者继续给予帮助和指正。

  本书由国家级特色专业建设项目、国家级教学团队建设项目、国家级精品课程建设项目、国家自然科学基金项目、中国气象局与南京信息工程大学局校共建的重点教材建设基金项目、江苏高校品牌专业建设工程资助项目(PPZY2015A016)及 2015 年江苏省高等教育教改研究立项课题(2015JSJG032)和南京信息工程大学"三百工程"建设精品教材建设项目资助,在此谨致诚挚谢忱。

作者

2019 年 2 月于南京

第二版前言

《气象科技英语教程》一书自出版以来，受到读者的广泛欢迎和好评，同时也得到了他们的很多宝贵意见和建议。值此再版的机会，我们将本书作了一次认真的修订。除寿绍文教授和姚永红博士外，寿亦萱博士对全书做了仔细的校订，沈新勇教授对部分词汇加注了音标。我们还请了加拿大专家玛格丽特·帕特森(Margaret Paterson)女士和苏珊娜·莫罗(Suzanne Moreau)女士及乔治·莫罗(Georges Moreau)先生对书中第二部分气象科技英语听说材料作了校订和朗读。我们希望通过这些工作使本书有所改进。在此，我们谨对所有给予我们关心、支持、帮助、指教的老师、同学和朋友们表示深切谢意，并希望能继续得到他们不断的批评指正。最后我们要特别说明本书中所有英文文章均源自国外材料，由于这些原文的出版者和作者的名字一时难以找到，所以暂时还不可能将他们一一列出，为此我们深表歉意。但是在此我们要对所有原文的外国出版者和作者们表示最衷心的感谢，正是由于他们的出色贡献和生动的文字使我们能够既学到了英语，又增长了气象的专业知识。

作者
2007 年 12 月
于南京信息工程大学

第一版前言

　　科技英语是描述科技用语中各种语言现象和特性的一种英语体系,它是用来进行国际科技交流的重要手段。全面的科技英语能力应包括读、写、译、听、说等五个主要方面。本教程主要为气象专业的学生在这五个主要方面的训练中提供必要的材料和知识。

　　《气象科技英语教程》共分三大部分:

　　第一部分为气象科技英语读物,是基本的科技英语阅读材料,内容包括各种文体的科技论文,一般都含有不少生词和词组以及长句、难句,主要训练阅读、翻译和写作能力。

　　第二部分为气象科技英语听说材料,这一部分都是一些篇幅短小、内容浅近、生动有趣的材料,主要训练听说能力。

　　第三部分为科技英语知识,目的是使学生对科技英语特点有较系统的认识。

　　由于气象科技英语课程学时有限,一般仅 50 学时左右,所以实际的课堂教学以第一部分为主,第二部分仅供课余自学,第三部分也以自学为主,教师可将这一部分的有关内容穿插在基本阅读材料的讲授过程中进行非常简要的介绍。本教程主要是在南京气象学院历届科技英语讲义基础上,参考有关文献并结合我们自己的教学体会编写而成的。在编写过程中得到许多老师多方面的关心、支持、帮助和指教。在此我们对他们表示深切谢意。最后,由于我们水平有限,时间仓促,书中错误和不足之处在所难免,敬请读者批评指正。

作者
2002 年 2 月

目 录

第三版前言
第二版前言
第一版前言

第一部分　气象科技英语读物
Part One: Reading Materials of English for Meteorological Science and Technology

1. The Structure and Composition of the Atmosphere ⋯⋯⋯⋯⋯⋯⋯⋯ (3)
2. Frontogenesis and Frontal Characteristics ⋯⋯⋯⋯⋯⋯⋯⋯⋯⋯⋯ (12)
3. Meteorological Forecasts ⋯⋯⋯⋯⋯⋯⋯⋯⋯⋯⋯⋯⋯⋯⋯⋯⋯⋯ (21)
4. The Greenhouse Effect ⋯⋯⋯⋯⋯⋯⋯⋯⋯⋯⋯⋯⋯⋯⋯⋯⋯⋯⋯ (29)
5. The General Circulation of the Atmosphere ⋯⋯⋯⋯⋯⋯⋯⋯⋯⋯⋯ (35)
6. Severe Storms ⋯⋯⋯⋯⋯⋯⋯⋯⋯⋯⋯⋯⋯⋯⋯⋯⋯⋯⋯⋯⋯⋯ (41)
7. Monsoon ⋯⋯⋯⋯⋯⋯⋯⋯⋯⋯⋯⋯⋯⋯⋯⋯⋯⋯⋯⋯⋯⋯⋯⋯ (51)
8. A History of Numerical Weather Prediction ⋯⋯⋯⋯⋯⋯⋯⋯⋯⋯⋯ (56)
9. Ozone ⋯⋯⋯⋯⋯⋯⋯⋯⋯⋯⋯⋯⋯⋯⋯⋯⋯⋯⋯⋯⋯⋯⋯⋯⋯ (63)
10. Weather Hazards in Agriculture ⋯⋯⋯⋯⋯⋯⋯⋯⋯⋯⋯⋯⋯⋯⋯ (71)
11. Polar-Orbiting Meteorological Satellites and Image Interpretation ⋯⋯⋯ (77)
12. Micrometeorology ⋯⋯⋯⋯⋯⋯⋯⋯⋯⋯⋯⋯⋯⋯⋯⋯⋯⋯⋯⋯ (84)
13. Approaches to Climatic Classification ⋯⋯⋯⋯⋯⋯⋯⋯⋯⋯⋯⋯⋯ (90)
14. Climatic Factors in Plant Growth ⋯⋯⋯⋯⋯⋯⋯⋯⋯⋯⋯⋯⋯⋯ (95)
15. The Tropical Ocean and Global Atmosphere Project ⋯⋯⋯⋯⋯⋯⋯⋯ (100)
16. GOES Data and Nowcasting ⋯⋯⋯⋯⋯⋯⋯⋯⋯⋯⋯⋯⋯⋯⋯⋯ (105)
17. Acidic Deposition ⋯⋯⋯⋯⋯⋯⋯⋯⋯⋯⋯⋯⋯⋯⋯⋯⋯⋯⋯⋯ (110)
18. Weather Modification ⋯⋯⋯⋯⋯⋯⋯⋯⋯⋯⋯⋯⋯⋯⋯⋯⋯⋯⋯ (115)
19. Agrometeorological Forecasting ⋯⋯⋯⋯⋯⋯⋯⋯⋯⋯⋯⋯⋯⋯⋯ (120)
20. Radar Measurement of Rainfall Intensity ⋯⋯⋯⋯⋯⋯⋯⋯⋯⋯⋯⋯ (125)
21. Jet Streams ⋯⋯⋯⋯⋯⋯⋯⋯⋯⋯⋯⋯⋯⋯⋯⋯⋯⋯⋯⋯⋯⋯ (132)
22. Meteorology ⋯⋯⋯⋯⋯⋯⋯⋯⋯⋯⋯⋯⋯⋯⋯⋯⋯⋯⋯⋯⋯⋯ (143)
23. Numerical Weather Prediction ⋯⋯⋯⋯⋯⋯⋯⋯⋯⋯⋯⋯⋯⋯⋯ (149)
24. Weather Forecasting and Its Accuracy ⋯⋯⋯⋯⋯⋯⋯⋯⋯⋯⋯⋯⋯ (155)

25. Long-Range Weather Prediction ……………………………………… (163)
26. Introducing the New Generation of Chinese Geostationary Weather Satellites, Fengyun-4 ……………………………………………………………… (170)

第二部分　气象科技英语听说材料
Part Two: Listening and Spoken English Materials for Meteorological Science and Technology

1. A Weather Report …………………………………………………… (183)
2. An Introduction to the U.S. Climate ………………………………… (184)
3. A U.S. Synoptic Chart ………………………………………………… (185)
4. An Area Weather Map ………………………………………………… (187)
5. Weather Forecasting in the U.S. ……………………………………… (188)
6. Meteorologists and Their Work ……………………………………… (189)
7. To Rain or Not to Rain ………………………………………………… (190)
8. Hail ……………………………………………………………………… (191)
9. Is Snow Coming? ……………………………………………………… (192)
10. Typhoons ……………………………………………………………… (193)
11. Hurricanes ……………………………………………………………… (195)
12. El Nino Phenomenon ………………………………………………… (197)
13. The Greenhouse Effect ……………………………………………… (198)
14. The Threat of Global Warming ……………………………………… (200)
15. Ozone …………………………………………………………………… (202)
16. The Hole over the Antarctic ………………………………………… (203)
17. How Solar Activity Affects Life on Earth …………………………… (205)

第三部分　气象科技英语知识
Part Three: Knowledge of English for Meteorological Science and Technology

1. 科技英语的特点 ……………………………………………………… (209)
2. 科技英语的词汇 ……………………………………………………… (210)
   2.1 专业词汇与普通词汇 …………………………………………… (210)
   2.2 生词的记忆 ……………………………………………………… (210)
   2.3 构词法 …………………………………………………………… (210)
3. 科技英语的语法 ……………………………………………………… (216)
   3.1 英语的基本句型 ………………………………………………… (216)

3.2　长句的分析 …………………………………………… (217)
4. 科技英语的习惯用语 ……………………………………… (219)
　　4.1　某些惯常的语法结构 ………………………………… (219)
　　4.2　词的习惯搭配 ………………………………………… (219)
5. 科技英语的翻译 …………………………………………… (236)
　　5.1　要准确选择词义 ……………………………………… (236)
　　5.2　可适当引申词义 ……………………………………… (236)
　　5.3　可适当增减词语 ……………………………………… (237)
　　5.4　可适当重复词语 ……………………………………… (237)
　　5.5　可适当改变句子成分或转换词性 …………………… (237)
　　5.6　被动语态的译法 ……………………………………… (238)
　　5.7　倒装句及后置修饰的译法 …………………………… (239)
　　5.8　定语从句的译法 ……………………………………… (239)
　　5.9　长难句的译法 ………………………………………… (240)
　　5.10　数量增减及倍数的译法 …………………………… (240)
　　5.11　近似值的译法 ……………………………………… (244)
　　5.12　数词短语的译法 …………………………………… (245)
　　5.13　不定数量的译法 …………………………………… (246)
　　5.14　量的尺度概念 ……………………………………… (247)
6. 数学用语的译、读 ………………………………………… (248)
7. 科技英语的写作 …………………………………………… (253)
　　7.1　科技论文的组成部分及其写作要求 ………………… (253)
　　7.2　有关写作的技术问题 ………………………………… (259)
8. 科技英语的快速阅读 ……………………………………… (264)
9. 学术报告和讨论 …………………………………………… (275)
　　9.1　开场白(点明主题) …………………………………… (275)
　　9.2　段落过渡 ……………………………………………… (275)
　　9.3　结束语 ………………………………………………… (276)
　　9.4　指要点 ………………………………………………… (276)
　　9.5　比较和对照 …………………………………………… (276)
　　9.6　以另一种方式重新阐述 ……………………………… (276)
　　9.7　增加理由、加强论证 ………………………………… (277)
　　9.8　举例子 ………………………………………………… (277)
　　9.9　指示图、表,并通过图、表谈论问题 ………………… (277)

  9.10 表示层次 ………………………………………………………… (278)
  9.11 下定义 …………………………………………………………… (278)
  9.12 做总结 …………………………………………………………… (279)
  9.13 讨论 ……………………………………………………………… (279)
10. 记笔记及缩略语 ……………………………………………………… (282)
  10.1 怎样记笔记 ……………………………………………………… (282)
  10.2 气象科学中常用的缩略语词汇 ………………………………… (284)
  10.3 缩略语的读音规则 ……………………………………………… (291)

主要参考文献 ……………………………………………………………… (292)

听说材料下载说明

  本书所用听说材料可以从气象出版社网站下载。网址为：http://www.qxcbs.com/ebook/qxkjyy/mdata.html。

第一部分 气象科技英语读物

Part One: Reading Materials of English for Meteorological Science and Technology

1. The Structure and Composition of the Atmosphere

Text

Like a fish in the ocean, man is confined to a very shallow layer of atmosphere. The gaseous envelope of the Earth is physically inhomogeneous in both the vertical and horizontal directions, although the horizontal inhomogeneity is much less marked than the vertical inhomogeneity.

Various criteria have been devised for dividing the atmosphere into layers. This division can be based on the nature of the vertical temperature profile, on the gaseous composition of the air at different altitudes, and the effect of the atmosphere on aircraft at different altitudes, etc. The division based on the variation of the air temperature with altitude is used most commonly in the meteorological literature.

According to a publication of the aerological commission of the World Meteorological Organization (WMO) in 1961, the Earth's atmosphere is divided into five main layers: the troposphere, the stratosphere, the mesosphere, the thermosphere and the exosphere. These layers are bounded by four thin transition regions: the tropopause, the stratopause, the mesopause and the thermopause.

The troposphere is the lower layer of the atmosphere between the Earth's surface and the tropopause. The temperature drops with increasing height in the troposphere, at a mean rate of 6.5℃ per kilometer (lapse rate). The upper boundary of the troposphere lies at a height of approximately 8 to 12 km in the polar and middle latitudes and 16 to 18 km in the tropics. In the polar and middle latitudes the troposphere contains about 75% of the total mass of atmospheric air, while in the tropics it contains about 90%. The tropopause is an intermediate layer in which either a temperature inversion or an isothermal temperature distribution is observed.

The stratosphere is the atmospheric layer above the troposphere. In the stratosphere the temperature either increases with height or remains nearly constant. In the lower part of the stratosphere (up to approximately 20 km above the Earth's surface), the temperature is practically constant (about $-56℃$). While further up the temperature increases with altitude at a rate of about 1℃/km at heights of 20 to 30 km and about 2.8℃/km at altitudes from 32 to 47 km. Under the standard conditions the temperature at the 47 km level is normally $-2.5℃$. This increase in temperature with height is due to the absorption of UV solar radiation by ozone molecules. It should be noted that about 99% of the total mass of atmospheric air is concentrated in the troposphere and stratosphere, which extend up to an altitude of 30 or 35 km. The stratopause is an intermediate layer between the stratosphere and the mesosphere (in the altitude region from 47 to 52 km), in which the temperature remains constant at about 0℃.

The mesosphere is an atmospheric layer in which the temperature continuously decreases with height at a rate of about 2.8℃/km up to about 71 km and at a rate of 2.0℃/km from 71 to 85 km. At heights of 85 to 95 km the temperature ranges from -85 to $-90℃$. The mesopause is an intermediate layer between the mesosphere and the thermosphere (the base of the temperature-inversion region in the thermosphere). Normally the mesopause has an altitude of 85 to 95 km and it is characterized by a constant temperature of about $-86.5℃$.

The thermosphere is the atmospheric layer above the mesopause. The temperature in this layer increases with increasing altitude, reaching about 2000 ℃ at about 450 km, the mean height of the upper boundary of the thermosphere. The temperature increase in this layer is mainly caused by the absorption of UV solar radiation by oxygen molecules, which dissociate as a result of this process.

The exosphere is the furthest out and the least studied part of the upper atmosphere. It is located above 450 km altitude. The air density in the exosphere is so low that atoms and molecules can escape from it into interplanetary space.

Finally, along with the above division of the atmosphere, we will also make use of a division based on the extent of atmospheric interaction with the Earth's surface. According to this principle, the atmosphere is usually divided into a so-called boundary layer (sometimes also called the friction layer) and the free atmosphere. The atmospheric boundary layer (up to 1 or 1.5 km) is influenced considerably by the Earth's surface and by eddy-viscosity forces. At the same time, we can neglect, as a first approximation, the influence of eddy-viscosity forces in the free atmosphere.

Of all the above atmospheric layers, only the troposphere (especially its boundary layer) is characterized by a marked instability of the vertical distribution of the meteorological parameters. It is in this layer that both temperature inversions and superadiabatic temperature variations with height are observed.

The Earth's atmosphere is a mixture of gases and aerosols, the latter being the name given to a system comprised of small liquid and solid particles distributed in the air. Air is not a specific gas; rather, it is a mixture of many gases. Some of them, such as nitrogen, oxygen, argon, neon, and so on, may be regarded as permanent atmospheric components that remain in fixed proportions to the total gas volume. Other constituents such as water vapor, carbon dioxide, and ozone vary in quantity from place to place and from time to time.

The principal sources of nitrogen, the most abundant constituent of air, are decaying from agricultural debris, animal matter and volcanic eruption. On the other side of the ledger, nitrogen is removed from the atmosphere by biological processes involving plants and sea life. To a lesser extent, lightning and high temperature combustion processes convert nitrogen gas to nitrogen compounds that are washed out of the atmosphere by rain or snow. The destruction of nitrogen is in the atmospheres in balance with production.

Oxygen, a gas crucial to life on Earth, has an average residence time in the atmosphere of about 3000 years. It is produced by vegetation that, in the photosynthetic growth process, takes up carbon dioxide and releases oxygen. It is removed from the atmosphere by humans and animals, whose respiratory systems are just the reverse of those of the plant communities. We inhale oxygen and exhale carbon dioxide. Oxygen dissolves in the lakes, rivers and oceans, where it serves to maintain marine organisms. It is also consumed in the process of decay of organic matter and in chemical reactions with many other substances. For example, the rusting of steel involves its oxidation.

From the human point of view, the scarce, highly variable gases are of great importance. The mass of water vapor, that is, H_2O in a gaseous state, in the atmosphere is relatively small and is added to and removed from the atmosphere relatively fast. As a result, the average residence time of water vapor is only 11 days. Water vapor is the source of rain and snow, without which we could not survive. From common experiences it is well known that the water vapor content of air varies a great deal. In a desert region the concentration of water vapor can be so low as to represent only a tiny fraction of the air volume. At the other extreme, in hot, moist air near sea level, say over an equatorial ocean, water vapor may account for as much as perhaps 5 percent of the air volume.

There are large variations of atmospheric water vapor from place to place and from time to time, but the total quantity over the entire Earth is virtually constant.

The same can not be said about carbon dioxide (CO_2). The concentration of this sparse but important gas has been increasing for the last hundred years or so. Carbon dioxide is added to the atmosphere by the decay of plant material and humus in the soil, and by the burning of fossil fuels: coal, oil and gas. The principal sinks of CO_2 are the oceans and plant life that uses CO_2 in photosynthesis. In the middle 1980s, atmospheric chemists were still debating about the effects on atmospheric CO_2 of burning, harvesting and clearing of forests. The oceans take up large amounts of CO_2, about half the amount released by fossil fuel combustion. It is expected that this fraction will diminish with the passing decades, whereas the total mass of CO_2 released will increase, at least through the early part of the next century. During the 1980s, atmospheric CO_2 was accumulating at a rate of about 1 part per million (ppm) of air per year, but it is expected to increase more rapidly in decades to come. In 1983 it averaged about 340 ppm of air.

Ozone (O_3), another important, highly variable gas, occurs mostly at upper altitudes, but it is also found in urban localities having a great deal of industry and automotive traffic and a generous supply of sunshine. In cities such as Los Angeles, ozone concentration may be more than 0.1 ppm in extreme cases. Most atmospheric ozone concentrations often exceed 1.0 ppm and may be as large as 10 ppm. They vary greatly with latitude, season, time of day and weather patterns. The high-altitude ozone layer is maintained by photochemical reactions. The ozone layer is important because, by absorbing UV radiation in the upper atmosphere, it reduces the amount reaching the surface of the Earth. Exposure to increased doses of ultraviolet rays would cause more severe sunburns and increase the risk of skin cancers. Biologists indicate that a substantial increase in UV radiation could also affect other components of the biosphere.

Certain gases, if they exist in sufficiently high concentrations, can be toxic to people, animal and plant life. For example, when ozone occurs in high concentrations, it is toxic to biological organisms. This does not happen often, but in heavily polluted localities such as Los Angeles, ozone near the ground sometimes is sufficiently abundant to cause leaf damage to certain plant species. Very large quantities of potentially hazardous gases are introduced into the atmosphere as a result of human activities. Air pollutants are emitted from furnaces, factories, refineries and engines, particularly automobile engines. All these things and others like them burn fossil fuels: coal, oil, gasoline and kerosene. In the process they emit gases and smoke particles that may spend a great deal of time in the atmosphere reacting with other substances and causing the formation of toxic compounds.

The most widespread and potentially hazardous gaseous pollutants are carbon monoxide, sulfur dioxide, nitrogen oxide and hydrocarbons. The last of these compounds comes from vaporized gasoline and other petroleum products.

New Words

composition	[kɔmpəˈziʃən]	n.	组成,成分
gaseous	[ˈgeizjəs]	a.	气体的
inhomogeneous	[ˌinhɔməˈdʒiːniəs]	a.	不均匀的
horizontal	[ˌhɔriˈzɔntl]	a.	水平的
inhomogeneity	[ˈinˌhəumədʒəˈniəti]	n.	不均匀性
marked	[maːkt]	a.	显著的
criterion (pl. criteria)	[kraiˈtiəriən]	n.	判断标准,依据
devise	[diˈvaiz]	v.	设计,发明,想出
profile	[ˈprəufail]	n.	分布,廓线
altitude	[ˈæltitjuːd]	n.	高度
meteorological	[ˌmiːtjərəˈlɔdʒikəl]	a.	气象的
literature	[ˈlitəritʃə]	n.	文献,著作
aerological	[ˌɛərələˈgeikəl]	a.	高空的
WMO (World Meteorological Organization)	[wɜːld][ˌmiːtjərəˈlɔdʒeikəl][ˌɔːgənaiˈzeiʃən]		世界气象组织
troposphere	[ˈtrɔpəusfiə]	n.	对流层
stratosphere	[ˈstrætəusfiə]	n.	平流层
mesosphere	[ˌmesəsfiə]	n.	中间层
thermosphere	[ˈθəːməsfiə]	n.	热成层
exosphere	[ˈeksəsfiə]	n.	外逸层
transition	[trænˈziʒən,-ˈsiʃən]	n.	过渡层
tropopause	[ˈtrɔpəupɔːz]	n.	对流层顶
stratopause	[ˈstrætəˌpɔːz]	n.	平流层顶
mesopause	[ˈmesəˌpɔːz, ˈmez-]	n.	中间层顶
thermopause	[ˈθəːməupɔːz]	n.	热成层顶
lapse rate	[læps][reit]	n.	递减率
boundary	[ˈbaundəri]	n.	边界
polar	[ˈpəulə]	a.	极的,极地的

latitude	[ˈlætitjuːd]	n.	纬度
tropics	[ˈtrɔpik]	n.	热带
atmospheric	[ˌætməsˈferik]	a.	大气的
inversion	[inˈvəːʃən]	n.	逆转,逆增(指温度),逆温,逆减(指降水量)
isothermal	[ˌaisəuˈθəːməl]	a.	等温的
distribution	[ˌdistriˈbjuːʃən]	n.	分布
absorption	[əbˈsɔːpʃən]	n.	吸收
ultraviolet (UV)	[ˈʌltrəˈvaiəlit]	a. n.	紫外线
ozone	[ˈəuzəun, əuˈzəun]	n.	臭氧
characterize	[ˈkæriktəraiz]	v.	表示…的特征,描述
dissociate	[diˈsəuʃieit]	v.	分解
density	[ˈdensiti]	n.	密度
interplanetary	[ˌintə(ː)ˈplænitəri]	a.	(行)星际的
interaction	[ˌintərˈækʃən]	n.	相互作用
friction	[ˈfrikʃən]	n.	摩擦
eddy-viscosity	[ˈedi-visˈkɔsiti]	n.	涡动黏滞性
instability	[ˌinstəˈbiliti]	n.	不稳定度
parameter	[pəˈræmitə]	n.	参数
superadiabatic	[ˈsjuːpəˌædiəˈbætik]	a.	超绝热的
aerosol	[ˈɛərəsɔl]	n.	气溶胶
nitrogen	[ˈnaitrədʒən]	n.	氮
argon	[ˈaːgɔn]	n.	氩
neon	[ˈniːən]	n.	氖
component	[kəmˈpəunənt]	n.	成分,分量,部分
constituent	[kənˈstitjuənt]	n.	成分
decay	[diˈkei]	v.	分解,腐败
agricultural	[ˌægriˈkʌltʃərəl]	a.	农业的
volcanic	[vɔlˈkænik]	a.	火山的
debris	[ˈdebriːiˈdeibriː]	n.	碎片
eruption	[iˈrʌpʃən]	n.	爆发,喷发
ledger	[ˈledʒe]	n.	账本
biological	[baiəˈlɔdʒikəl]	a.	生物的
destruction	[disˈtrʌkʃən]	n.	破坏,毁灭
residence	[ˈrezidəns]	n.	驻留,存在
crucial	[ˈkruːʃəl]	a.	极重要的,决定性的

photosynthetic	[ˌfəutəusin'θetik]	a. 光合的
respiratory	[ris'paiərətəri]	a. 呼吸的
inhale	[in'heil]	v. 吸入
exhale	[eks'heil]	v. 呼出
dissolve	[di'zɔlv]	v. 溶解,分解
marine	[mə'ri:n]	a. 海洋的
organism	['ɔ:gənizəm]	n. 有机体
organic	[ɔ:'gænik]	a. 有机的,器官的
rust	[rʌst]	v. n. 生锈
oxidation	[ɔksi'deiʃən]	n. 氧化[作用]
concentration	[ˌkɔnsen'treiʃen]	n. 浓度
equatorial	[ˌekwə'tɔ:riəl]	a. 赤道的
virtually	['və:tjuəli]	ad. 实际上
sparse	[spa:s][skɛəs]	a. 稀少的(scarce)
humus	['hju:məs]	n. 腐殖质
fossil	['fɔsl]	n. 化石
sink	[siŋk]	n. v. 汇
debate	[di'beit]	v. 争论,辩论
diminish	[di'miniʃ]	v. 减少,减小
ppm (part per million)	[pa:t][pə:]['miljən]	百万分之一
toxic	['tɔksik]	a. 有毒的
absorber	[əb'sɔ:bə]	n. 吸收体
infrared	['infrə'red]	n. a. 红外线
emit	[i'mit]	v. 发射
locality	[ləu'kæliti]	n. 地区,位置
generous	['dʒenərəs]	a. 丰富的,慷慨的
photochemical	[ˌfəutəu'kemikəl]	a. 光化学的
sunburn	['sʌnbə:n]	n. 太阳灼伤
substantial	[səb'stænʃəl]	a. 显著的,实质的,基本上的
biosphere	['baiəsfiə]	n. 生物圈
pollutant	[pə'lu:tənt]	n. 污染物
refinery	[ri'fainəri]	n. 提炼厂,炼油厂
smelter	['smeltə]	n. 冶炼厂
gasoline	['gæsəli:n]	n. 汽油
kerosene	['kerəsi:n]	n. 煤油
widespread	['waidspred;-'spred]	a. 广泛的

potentially	[pə'tenʃəli]	ad. 潜在的
hazardous	['hæzədəs]	a. 有危害的
sulfur	['sʌlfə]	n. 硫
hydrocarbon	['haidrəu'kɑːbən]	n. 碳氢化合物,烃类
petroleum	[pi'trəuliəm]	n. 石油

Exercises (Reviewing and Thinking)

1. Answer the following questions according to the text by oral or writing

(1) What is the Earth atmosphere?

(2) How to divide the atmosphere into layers?

(3) Which criterion is most commonly used for layer's division?

(4) What is the troposphere?

(5) What is the stratosphere?

(6) How the vertical lapse rate of temperature changes with height?

(7) Why the temperature increases with height in stratosphere?

(8) What is the mesosphere?

(9) How high is the mesosphere located above the Earth's surface?

(10) How the temperature changes with height in the mesosphere?

(11) How high is the temperature on the mesopause?

(12) What is the thermosphere?

(13) What is the exosphere?

(14) What gases may be regarded as permanent atmospheric components?

2. Homework

(1) To make a table to list the composition of the Earth's atmosphere according to the text

	Sources	Destruction	Residence time
Nitrogen			
Oxgen			
H_2O			
CO_2			

(2) According to the following picture (Fig. 1) to prepare a speech talking about the atmospheric layers and the vertical profile of temperature.

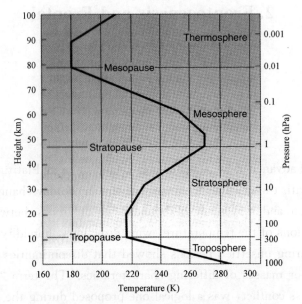

Fig. 1　The schematic diagram of the atmospheric structure

Supplemental Reading

The Basis for Dividing Atmospheric Layers

The vertical distribution of temperature for typical conditions in the Earth's atmosphere, shown in Fig. 1, provides a basis for dividing atmosphere into four layers (troposphere, stratosphere, mesosphere, and thermosphere), the upper limits of which are denoted by the suffix pause.

The troposphere (turning or changing) is marked by generally decreasing temperatures with height, at an average lapse rate, of $-6.5℃\ \mathrm{km}^{-1}$. That is to say,

$$\Gamma \equiv -\frac{\partial T}{\partial z} \sim 6.5℃ \cdot \mathrm{km}^{-1} = 0.0065℃ \cdot \mathrm{m}^{-1}$$

Where T is temperature and Γ is the lapse rate.

(Adapted from Wallace and Hobbs, 2006)

2. Frontogenesis and Frontal Characteristics

Text

Frontogenesis

The first real advance in our detailed understanding of mid-latitude weather variations was made with the discover that many of the day-to-day changes are associated with the formation and movement of boundaries, or fronts, between different air masses. Observations of the temperature, wind directions, humidity and other physical phenomena during unsettled periods showed that discontinuities often persist between impinging air masses of differing characteristics. The term "front", for these surfaces of air-mass conflict, was a logical one proposed during the First World War by a group of meteorologists working in Norway, and their ideas are still an integral part of most weather analysis and forecasting particularly in middle and high latitudes.

1. Frontal waves

It was observed that the typical geometry of the air-mass interface, or front, resembles a wave form. Similar wave patterns are, in fact, found to occur on the interface between many different media, for example, waves on sea surface, ripples on beach sand, aeolian sand dunes, etc. Unlike these wave forms, however, the frontal waves in the atmosphere are commonly unstable: that is, they suddenly originate, increase in size, and then gradually dissipate. Numerical model calculations show that, in middle latitudes waves in a baroclinic atmosphere are unstable if their wavelength exceeds a few thousand kilometers. Frontal wave cyclones are typically 1500—3000 km in wavelength. The initially attractive analogy between atmospheric wave systems and waves formed on interface of other media is, therefore, an insufficient-basis on which to develop explanations of frontal waves. In particular, the circulation of the upper troposphere plays a key role in providing appropriate conditions for their development and growth, as will be shown below.

2. The frontal wave depression

A depression (also termed a low or cyclone) is an area of relatively low pressure, with a more or less circular isobaric pattern. It covers an area 100—3000 km in diameter and usually has a life-span of 4—7 days. Systems with these characteristics, which are prominent on daily weather maps are referred to as synoptic scale features. The depression, in mid-latitudes at least, is usually associated with a convergence of contrasting air masses. The interface between these air masses develops into a wave form with its apex located at the centre of the low-pressure area. The wave encloses a mass of warm air between modified cold air in front and fresh cold air in the rear. The formation of the wave also creates a distinction between the two sections of the original air-mass discontinuity for, although each section still marks the boundary between cold and warm air, the weather characteristics found in the neighborhood of each section are very different. The two sections of the frontal surface are distinguished by the names: warm front for the leading edge of the wave, and cold front for that of the cold air to the rear.

The depression usually achieves its maximum intensity 12—24 hours after the beginning of occlusion.

Frontal Characteristics

The activity of a front in terms of weather depends upon the vertical motion in the air masses. If the air in the warm sector is rising relative to the frontal zone the fronts are usually very active and are termed ana-fronts. Whereas sinking of the warm air relative to the cold air masses gives rise to less inactive kata-fronts.

1. The warm front

The warm front represents the leading edge of the warm sector in the wave. The frontal zone here has a very gentle slope, of the order $\frac{1}{2}°-1°$, so that the cloud systems associated with the upper portion of the front herald its approach some 12 hours or more before the arrival of the surface front. The ana-warm front, with rising warm air, has multi-layered cloud which steadily thickens and lowers towards the surface position of the front. The first clouds are thin, wispy cirrus, followed by sheets of cirrus and cirrostratus, and altostratus. The sun is obscured as the altostratus layer thickens, and drizzle or rain begins to fall. The cloud often extends through most of the troposphere and with continuous precipitation occurring is generally designated as nimbostratus. Patches of stratus may also form in the cold air as rain falling through this air undergoes evaporation and quickly saturates it.

The descending warm air of the kata-warm front greatly restricts the development of medium- and high-level clouds. The frontal cloud is mainly stratocumulus, with a limited depth as a result of the subsidence inversions in both air masses. Precipitation is usually light rain or drizzle formed by coalescence since the freezing level tends to be above the inversion layer, particularly in summer.

In the passage of the warm front the wind veers, the temperature rises and the fall of pressure is checked. The rain becomes intermittent or ceases in the warm air and the thin stratocumulus cloud sheet may break up.

Forecasting the extent of rain belts associated with the warm front is complicated by the fact that most fronts are not ana- or kata-fronts throughout their length or even at all levels in the troposphere. For this reason, radar is being used increasingly to determine by direct means the precise extent of rain belts and to detect differences in rainfall intensity.

Such studies have shown that most of the production and distribution of precipitation are controlled by a broad airflow a few hundred kilometres across and several kilometers deep, which flows parallel to and ahead of the surface cold front.

Just ahead of the cold front the flow occurs as a low-level jet with winds up to $25-30 \text{ m} \cdot \text{s}^{-1}$ at about 1 km above the surface. The air, which is warm and moist, rises over the warm front and turns southeastward ahead of it as it merges with the midtropospheric flow. This flow has been termed a "conveyor belt" (for large-scale heat and momentum transfer in mid-latitudes). Broad-scale convective (potential) instability is generated by the over-running of this low-level flow by potentially colder, drier air in the middle troposphere. Instability is released mainly in small-scale convection cells that are organized into clusters, known as mesoscale precipitation areas (MPAs). These MPAs are further arranged in bands, 50—100 km wide. Ahead of the warm front, the bands are broadly parallel to the airflow in the rising section of the conveyor belt, whereas in the warm sector they parallel the cold front and the low-level jet. In some cases, cells and clusters are further arranged in bands within the warm sector and ahead of the warm front. Precipitation from warm front rainbands often involves "seeding" by ice particles falling from the upper cloud layers. It has been estimated that 20%—35% of the precipitation originates in the "seeder" zone and the remainder in the lower clouds. Some of the cells and clusters are undoubtedly set up through orographic effects and these influences may extend well down-wind when the atmosphere is unstable.

2. The cold front

The weather conditions observed at cold fronts are equally variable, depending upon the stability of the warm sector air and the vertical motion relative to the frontal zone. The classical cold-front model is of the ana-type, and the cloud is usually cumulonimbus. Over the British Isles air in the warm sector is rarely unstable, so that nimbostratus occurs more frequently at the cold front. With the kata-cold front the cloud is generally stratocumulus and precipitation is light. With ana-cold fronts there are usually brief, heavy downpours sometimes accompanied by thunder. The steep slope of the cold front, roughly 2°, means that the bad weather is of shorter duration than at the warm front. With the passage of the cold front the wind veers sharply, pressure begins to rise and temperature falls. The sky may clear very abruptly, even before the passage of the surface cold front in some cases, although with kata-cold fronts the changes are altogether more gradual.

3. The occlusion

Occlusions are classified as either cold or warm, the difference depending on the relative states of the cold air masses lying in front and to the rear of the warm sector. If the air is colder than the air following it then the occlusion is warm, but if the reverse is so, it is termed a cold occlusion. The air in advance of the depression is most likely to be coldest when depression occlude over Europe in winter and very cold of air is affecting the continent.

The line of the warm air wedge aloft is associated with a zone of layered cloud and often of precipitation. Hence its position is indicated separately on some weather maps and it is referred to by Canadian meteorologists as a trowal. The passage is an occluded front and trowal brings a change back to polar air-mass weather.

A different process occurs when there is interaction between a polar trough and the main polar front, giving rise to an instant occlusion. A warm conveyor belt on the polar front ascends an upper tropospheric jet forming a stratiform cloud band, while a low-level polar trough conveyor belt at right angles to it produces a convective cloud band and precipitation area poleward of the main polar front on the leading edge of the cold pool.

The occurrence of frontolysis (frontal decay) is not necessarily linked with occlusion, although it represents the final phase of a front's existence. Decay occurs when differences no longer exist between adjacent air masses. This may arise in four ways: through their mutual stagnation over a similar surface, as a result of both air masses moving on parallel tracks at the same speed, as a result of their movement in succession along the same track at the same speed, or by the system incorporating into itself air of the same temperature.

New Words

frontogenesis (frontogeneses)	[ˌfrʌtəu'dʒenisis]	n.	锋生
frontal	['frʌntl]	a.	锋(面)的
mid-latitude	[mid-'lætitjuːd]	n.	中纬度
day-to-day	[dei-tu-dei]	a.	逐日的
formation	[fɔː'meiʃən]	n.	形成
front	[frʌnt]	n.	锋,锋面
mass	[mæs]	n.	团,块
humidity	[hjuː'miditi]	n.	湿度
unsettled	['ʌn'setld]	a.	未解决的,未确定的,不安全的,不稳定的
discontinuity	['disˌkɔnti'njuː(ː)iti]	n.	不连续性
persist	[pə(ː)'sist]	v.	持续,坚持
impinging	[im'pindʒiŋ]	a.	紧密接触的
conflict	['kɔnflikt]	n.	交会
meteorologist	[ˌmiːtjə'rɔlədʒist]	n.	气象学家
Norway	['nɔːwei]	n.	挪威
integral	['intigrəl]	a.	完整的,整数的,积分的,必备的,必要的,主要的
interface	['intə(ː)ˌfeis]	n.	交面,接口
medium (media)	['miːdjəm] ['miːdjə]	n.	介质
ripple	['ripl]	n.	波纹,涟漪
aeolian	[iː'əuljən]	a.	风成的
dune	[djuːn]	n.	沙丘
dissipate	['disipeit]	v.	消散
numerical	[njuː(ː)'merikəl]	a.	数值的
model	['mɔdl]	n.	模式
baroclinic	[ˌbærə'klinik]	a.	斜压的

cyclone	[ˈsaikləun]	n.	气旋
attractive	[əˈtræktiv]	a.	有吸引力的
depression	[diˈpreʃən]	n.	低压
low	[ləu]	n.	低压,低压区
isobaric	[ˌaisəuˈbærik]	a.	等压的
span	[spæn]	n.	跨度,延伸
prominent	[ˈprɔminənt]	a.	突出的
synoptic	[siˈnɔptik]	a.	天气的
convergence	[kənˈvɜːdʒəns]	n.	辐合
contrasting	[kənˈtrɑːstiŋ]	a.	成对比的,相对立的
apex	[ˈeipeks]	n.	顶点,峰尖
modify	[ˈmɔdifai]	v.	修饰
rear	[riə]	n.	尾部
distinction	[disˈtiŋkʃən]	n.	差别,不同
neighbourhood	[ˈneibəhud]	n.	邻区
mark	[mɑːk]	v.	标志,标出
zone	[zəun]	n.	区域
cirrus (cirri)	[ˈsirəs]	n.	卷云
sector	[ˈsektə]	n.	区段
ana-front	[ˈɑːnə-frʌnt]	n.	上滑锋
kata-front	[ˈkɑːta-frʌnt]	n.	下滑锋
gentle	[ˈdʒentl]	a.	轻微的
slope	[sləup]	n.	倾斜,斜坡
order	[ˈɔːdə]	n.	量级
herald	[ˈherəld]	v.	预示…的来临,预报…的到来
multi-layered	[ˈmʌlti-ˈleiəd]	a.	多层的
wispy	[ˈwispi]	a.	稀疏的,缥缈的
cirrostratus (cirrostrati)	[ˈsirəuˈstrɑːtəs]	n.	卷层云
altostratus (altostrati)	[ˈæltəuˈstreitəs]	n.	高层云
obscure	[əbˈskjuə]	v.	遮蔽,使暗淡无光
drizzle	[ˈdrizl]	n.	毛毛雨
precipitation	[priˌsipiˈteiʃən]	n.	降水
designate	[ˈdezigneit]	v.	指定,把…叫做
nimbostratus (nimbostrati)	[ˈnimbəuˈstreitəs;-ˈstrætəs]	n.	雨层云
stratus (strati)	[ˈstreitəs]	n.	层云
saturate	[ˈsætʃəreit]	v.	使饱和
descend	[diˈsend]	v.	下降

medium-level	[ˈmiːdjəm-ˈlev(ə)l]	n.	中层
high-level	[hai-ˈlev(ə)l]	n.	高层
stratocumulus (stratocumuli)	[ˌstrætəuˈkjuːmjuləs]	n.	层积云
coalescence	[ˌkəuəˈlesns]	n.	(雨滴的)合并
veer	[viə]	v.	转向
airflow	[ˈeəfləu]	n.	气流
merge	[məːdʒ]	v.	合并
mid-tropospheric	[mid-trɔpəuˈsferik]	a.	对流层中部的
conveyor	[kənˈveiə]	n.	传送
large-scale	[laːdʒ-skeil]	a. n.	大尺度的
momentum (momenta)	[məuˈmentəm]	n.	动量
broad-scale	[brɔːd-skeil]	a.	大范围的
convective	[kənˈvektiv]	a.	对流性的
potential	[pəˈtenʃ(ə)l]	a.	位势(米),潜在的,潜力
cell	[sel]	n.	单体,单元
cluster	[ˈklʌstə]	n.	群,串
mesoscale	[ˈmesəu-skeil]	a. n.	中尺度(的)
rainband	[ˈreinbænd]	n.	雨带
seed	[siːd]	v.	撒播
seeder	[ˈsiːdə]	n.	催化剂
remainder	[riˈmeində]	n.	其余,余数,余项
orographic	[ˈɔrəgræfik]	a.	地形的,山形的
down-wind	[daun-waind]	ad.	在下风方向
ana-type	[ˈaːnə-taip]	n.	上滑型
cumulonimbus (cumulonibi)	[ˈkjuːmjuləuˈnimbəs]	n.	积雨云
isle	[ail]	n.	小岛
downpour	[ˈdaunpɔː]	n.	倾盆大雨
thunder	[ˈθʌndə]	n.	雷
duration	[djuəˈreiʃən]	n.	持续(时间),持久
occlude	[əˈkluːd]	v.	锢囚
trowal (trough of warm air aloft)	[trəˈwɔl]	n.	高空暖舌
aloft	[əˈlɔft]	ad. a.	高,上
ascend	[əˈsend]	v.	上升
stratiform	[ˈstrætiˌfɔːm]	n.	层状
poleward	[ˈpəulwəd]	a.	向极(地)的
frontolysis (frontolyses)	[frʌnˈtɔlisis]	n.	锋消
decay	[diˈkei]	n.	消亡

phase	[feiz]	n. 相位,阶段,方面
stagnation	[stæg'neiʃən]	n. 停滞
succession	[sək'seʃən]	n. 继续
incorporate	[in'kɔːpəreit]	v. 加入,合并

Exercises (Reviewing and Thinking)

1. Answer the following questions according to the text by oral or writing

(1) What are the definitions of cold front, warm front and stationary front respectively?

(2) What are the cold and warm occlusion fronts respectively?

(3) What are the definitions of depression and cyclone respectively?

(4) What are the definitions of frontogenesis and frontolysis respectively?

2. Homework

(1) Translate the following paragraph into Chinese.

Just ahead of the cold front the flow occurs as a low level jet with winds up to $25-30 \text{m} \cdot \text{s}^{-1}$ at about 1 km above the surface. The air, which is warm and moist, rises over the warm front and turns southeastward ahead of it as it merges with the mid-tropospheric flow. This flow has been termed a "conveyor belt" (for large scale heat and momentum transfer in mid-latitudes). Broad scale convective (potential) instability is generated by the over running of this low level flow by potentially colder, drier air in the middle troposphere. Instability is released mainly in small scale convection cells that are organized into clusters, known as mesoscale precipitation areas (MPAs). These MPAs are further arranged in bands, 50—100 km wide. Ahead of the warm front, the bands are broadly parallel to the airflow in the rising section of the conveyor belt, whereas in the warm sector they parallel the cold front and the low level jet. In some cases, cells and clusters are further arranged in bands within the warm sector and ahead of the warm front. Precipitation from warm front rainbands often involves "seeding" by ice particles falling from the upper cloud layers. It has been estimated that 20%—35% of the precipitation originates in the "seeder" zone and the remainder in the lower clouds. Some of the cells and clusters are undoubtedly set up through orographic effects and these influences may extend well down wind when the atmosphere is unstable.

(2) To prepare a oral presentation to talk about the structure of a cyclone based

on the illustration pictures in Fig. 1.

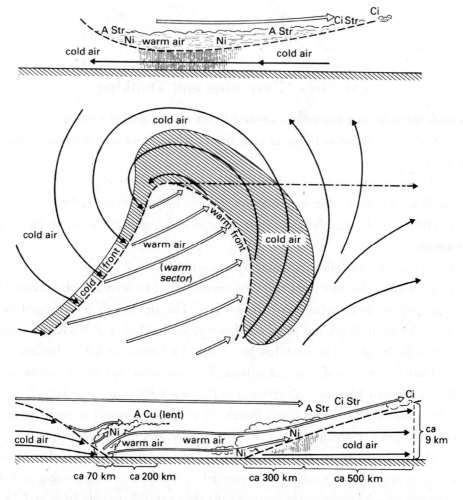

Fig. 1 The structure of a cyclone. (top: the vertical section crossing through over the wave apex; middle: the surface structure of the cyclone; bottom: the vertical section crossing over the cold and warm fronts and the warm sector of the cyclone)

3. Meteorological Forecasts

Text

National Meteorological Services perform a variety of activities in order to provide weather forecasts. The principal ones are data collection, the preparation of basic analyses and prognostic charts of atmosphere conditions for use by local weather offices, the preparation of short- and long-term forecasts for the public as well as special services for aviation, shipping, agricultural and other commercial and industrial users, and the issuance of severe weather warnings.

1. Data sources

The data required for forecasting and other services are provided by worldwide standard synoptic reports at 00, 06, 12 and 18 GMT, similar observations made hourly, particularly in support of national aviation requirements, upper-air soundings (at 00 and 12 GMT), satellite data and other specialized networks such as radar stations for severe weather. Under the World Weather Watch (WWW) program, synoptic reports are made at some 4,000 land stations and by 7,000 ships. There are about 700 stations making upper-air soundings (temperature, pressure, humidity and wind). These data are transmitted in code via teletype and radio links in regional or national centers and into the high-speed Global Telecommunications System (GTS) connecting World Weather Centers in Melbourne, Moscow and Washington and eleven Regional Meteorological Centers for redistribution. Some 157 states and territories (editor's note: according to new data, the number of states and territories of WMO is 191 before January 1st 2013) cooperate in this activity under the aegis of the World Meteorological Organization(WMO).

Meteorological information has been collected operationally by satellites of the United States and USSR since 1965 and, more recently, by the European Space Agency, India and Japan. There are two general categories of weather satellite: polar orbiters providing global coverage twice per 24 hours in orbital strips over the poles (such as the United States NOAA and TIROS series, and the USSR's Meteor) and geosynchronous satellites (such as the Geostationary Operational Environmental Satellites (GOES) and Metosat), giving repetitive (30-minute) coverage of almost one third of the earth's surface in low middle latitudes. Information on the atmosphere is collected as digital data or direct readout visible and infrared images of cloud cover and sea-surface temperature, but also includes global temperature and moisture profiles through the atmosphere obtained from multi-channel infrared and microwave sensors which receive radiation emitted from particular levels in the atmosphere. Additionally, satellites have a data collection system (DCS) that relays data on numerous environmental variations from ground platforms or ocean buoys to processing centers; GOES can also transmit processed satellite images in facsimile and the NOAA polar orbiters have an automatic picture transmission (APT) system that is utilized at 900 stations worldwide.

2. Forecasting

Modern forecasting did not become possible until weather information could be rapidly collected, assembled and processed. The first development came in the middle of the last century with the invention of telegraphy, which permitted immediate analysis of weather data by the drawing of synoptic charts. These were first displayed in Britain at the Great Exhibition of 1851. Sequences of weather change were correlated with barometric pressure patterns both in space and time by such workers as Fitzroy and Abereroleby, but it was not until later that theoretical models of weather systems were devised——notably the Bjerknes depression model.

Forecasts are usually referred to as short-range, medium (or extended) range and long-range. The first two can for present purposes be considered together.

Short-Range Forecasting

Forecasting procedures developed up to the 1950s were based on synoptic principles but, since the 1960s, practices have been revolutionized by numerical forecasting models and the adoption of "nowcasting" techniques.

During the first half of the 20th century, short-range forecasts were based on synoptic principles, empirical rules and extrapolation of pressure changes.

Since 1955 routine forecasts have been based on numerical models. These predict the evolution of physical processes in the atmosphere by determinations of the conservation of mass, energy and momentum. The basic principle is that the rise or fall of surface pressure is related to mass convergence or divergence, respectively, in the overlying air column.

Forecast practices in the major national centers are basically similar. The forecasts are essentially derived from twice-daily (00 and 12 GMT) prognoses of atmospheric circulation. Since most techniques are now largely automated, the analyses of synoptic fields are based on the previous 12-hour forecast maps as a first guess. Three different interpolation methods are used to obtain smoothed, grided data on temperature, moisture, wind and geopotential height for the surface at standard pressure levels (850,700,500,400,300,250,200 and 100 hPa) over the globe. The NMC currently has two basic prediction models: a special model with (6 or) 12 layers (from the boundary layer into the upper stratosphere), which is integrated for up to 10 days, and a regionally applicable nested grid model with finer horizontal resolution. It should be noted that typically the computer time required increases several-fold when the grid spacing is halved. The essential forecast products are MSL pressure, temperature and wind velocity for standard pressure levels, 1000—500 hPa thickness, vertical motion and moisture content in the lower troposphere, and precipitation amounts.

Actual weather conditions are now commonly predicted using the Model Output Statistics (MOS) technique developed by the US National Weather Service. Rather than relating weather variables to the predicted pressure/height patterns and taking account of frontal models, for example, a series of regression equations are developed for specific locations between the variable of interest and up to 10 predictors calculated by the numerical models. Weather elements so predicted for numerous locations include daily maximum/minimum temperature, 12-hour probability of precipitation occurrences and precipitation amount, probability of frozen precipitation, thunderstorm occurrence, cloud cover and surface winds. These forecasts are distributed as facsimile maps and tables to weather offices for local use.

Errors in numerical forecasts arise from several sources. One of the most serious is the limited accuracy of the initial analyses due to data deficiencies. The average over the oceans is sparse and only a quarter of the possible ship reports may be received within 12 hours; even over land more than one-third of the synoptic reports may be delayed beyond 6 hours. However, satellite-derived information and aircraft reports can help fill some gaps for the upper air. Another limitation is imposed by the horizontal and vertical resolution of the models and the need to parameterize sub-grid processes such as cumulus convection. The small-scale nature of the turbulent motion of the atmosphere means that some weather phenomena are basically unpredictable, for example, the specific locations of shower cells in an unstable airmass. Greater precision that the "showers and bright periods" or "scattered showers" of the forecast language is impossible with present techniques. The procedure for preparing a forecast is becoming much less subjective, although in complex weather situations the skill of the experienced forecaster still makes the technique almost as much as art as a science. Detailed regional or local predictions can only be made within the framework of the general forecast situation for the country and demand thorough knowledge of possible topographic or other local effects by the forecaster.

Nowcasting

Severe weather is typically short-lived (<2 hr) and, due to its mesoscale character (<100 km), it affects local/regional areas necessitating site-specific forecasts. Included in this category are thunderstorms, gust fronts, tornadoes, high winds especially along coasts, over lakes and mountains, heavy snow and freezing precipitation. The development of radar networks, now instruments and high-speed communication links has provided a means of issuing warnings of such phenomena. Several countries have recently developed integrated satellite and radar systems to provide information on the horizontal and vertical extent of thunderstorms, for example. Such data are supplemented by networks of automatic weather stations (including buoys) that measure wind, temperature and humidity. In addition, for detailed boundary layer and lower troposphere data, there is now an array of vertical sounders——acoustic sounders (measuring wind speed and direction from echoes created by thermal eddies), specialized (Doppler) radar measuring winds in clear air by returns either from insects (3.5 cm wavelength radar) or from variations in the air's refractive index (10 cm wavelength radar). Nowcasting techniques use highly automated computers and image analysis systems to integrate data from a variety of sources rapidly. Interpretation of the data displays requires skilled personnel and/or extensive software to provide appropriate information. The prompt forecasting of wind shear and downburst hazards at airports is one example of the importance of nowcasting procedures.

Overall, the greatest benefits from improved forecasting can be expected in aviation and the electric power industry for forecast less than 6 hours ahead, in transpiration, construction and manufacturing for 12—24 hour forecasts and in agriculture for 2—5 day forecasts. In terms of economic losses, the last category could benefit the most from more reliable and more precise forecasts.

Long-Range Forecasting

The methods discussed above are unsuitable for predicting the probable trend of the weather for periods of a month or more, because they are concerned with individual synoptic disturbances with a life cycle of about 3 to 7 days. Theoretical considerations indicate that the limit of synoptic predictability using numerical techniques is less than 15 days. Two rather different approaches will be dealt with later.

New Words

aviation	[ˌeiviˈeiʃən]	n. 飞行,航空
issuance	[ˈiʃu(:)əns]	n. 发布
warning	[ˈwɔːniŋ]	n. 警报
worldwide	[ˈwɜːldwaid]	a. 世界范围的
GMT (Greenwich Mean Time)	[ˈgrinidʒ][miːn][taim]	格林尼治标准时间
upper-air	[ˈʌpə-ɛə]	a. 高空的
sounding	[ˈsaundiŋ]	n. 探测
network	[ˈnetwəːk]	n. 网状组织
watch	[wɔtʃ]	n. 监视
WWW (World Weather Watch)		世界天气监视网
code	[kəud]	n. v. 电码,编码
teletype	[ˈteliˌtaip]	v. 电传打字机
telecommunication	[ˈtelikəmjuːniˈkeiʃən]	n. 无线电远程通讯
GTS (global telecommunication system)		全球电(传通)信系统
Melbourne	[ˈmelbən]	n. (澳)墨尔本
Moscow	[ˈmɔskəu]	n. (苏)莫斯科
redistribution	[ˈriːdistriˈbjuːʃən]	n. 再分发
territory	[ˈteritəri]	n. 地区,领域
cooperate	[kəuˈɔpəreit]	v. 合作
aegis	[ˈiːdʒis]	n. 保护
under the aegis of		在…支持下
operationally	[ˌɔpəˈreiʃənəli]	ad. 业务上
USSR (Union of Soviet Socialist Republics)	[ˈjuːnjən][ˈsəuviət][ˈsəuʃəlist] [riˈpʌblik]	苏联
agency	[ˈeidʒənsi]	n. 代办处,机构
European Space Agency		欧洲空间管理局
orbiter	[ˈɔːbitə]	n. 轨道飞行器
coverage	[ˈkʌvəridʒ]	n. 覆盖

strip	[strip]	n. 长带
NOAA (National Oceanic and Atmospheric Administration)	[ˈnæʃənəl][ˌəuʃiˈænik][ˌætməsˈferik][ədminisˈtreiʃən]	诺阿卫星(美国国家海洋大气管理局)
TIROS (television and infrared observing satellite)	[ˈteliviʒən][ˈinfrəred][əbˈzəːviŋ][ˈsætəlait]	泰罗斯卫星(电视红外业务卫星)
Meteor	[ˈmiːtjə]	n. 流星卫星
geosynchronous	[ˌdʒiːəuˈsiŋkrənəs]	a. 地球同步的
geostationary	[ˌdʒi(ː)əuˈsteiʃənəri]	a. 相对于地球静止的
GOES (geostationary operational environmental satellite)	[ˌɔpəˈreiʃnl][inˌvaiərənˈmentl]	地球静止业务环境卫星
Metosat		n. 梅多沙特卫星
repetitive	[riˈpetitiv]	a. 重复的
digital	[ˈdidʒitl]	a. 数字的
readout	[ˈriːdaut]	n. 读出
multi-channel	[ˈmʌlti-ˈtʃənl]	a. 多通道的
microwave	[ˈmaikrəuweiv]	n. 微波
sensor	[ˈsensə]	n. 感应器
DCS (data collection system)	[deitə][kəˈlekʃən]	资料收集系统
relay	[ˈriːlei]	v. 中继
buoy	[bɔi]	n. 浮标站
process	[prəˈses]	v. 加工,处理,整理
facsimile	[fækˈsimili]	n. 传真
APT (automatic picture transmission)	[ˌɔːtəˈmætik][ˈpiktʃə][ˈtrænzˈmiʃən]	自动图片发送
assemble	[əˈsembl]	v. 组装,汇编
telegraphy	[tiˈlegrəfi]	n. 电报
Great Exhibition	[greit][ˌeksiˈbiʃən]	博览会
barometric	[ˌbærəuˈmetrik]	a. 气压的
short-range	[ʃɔː-reindʒ]	a. 短期的
extend	[iksˈtend]	v. 延伸
long-range	[lɔŋ-reindʒ]	a. 长期的
revolutionize	[ˌrevəˈl(j)uːʃənaiz]	v. (使)革命化,重大改变
nowcasting	[nauˈkɑːstiŋ]	n. 现时预报
empirical	[emˈpirikəl]	a. 经验的
extrapolation	[ˌekstrəpəuˈleiʃən]	n. 外推
forecaster	[ˈfɔːkɑːstə]	n. 预报员
United Kingdom	[juˈnaitid][ˈkiŋdəm]	英国

routine	[ruˈtiːn]	n.	常规
conservation	[ˌkɔnsəˈveiʃən]	n.	保守
divergence	[daiˈvɜːdʒəns]	n.	辐散
derive	[diˈraiv]	v.	导出，获得
necessitate	[niˈsesiteit]	v.	使…成为需要，以…为条件
prognosis (prognoses)	[prɔgˈnəusis]	n.	预测
interpolation	[inˌtəːpəuˈleiʃən]	n.	内插
smooth	[smuːð]	v.	弄平滑，a. 平滑的
geopotential	[ˌdʒiːəupəˈtenʃəl]	a.	位势的
prediction	[priˈdikʃən]	n.	预报
spectral	[ˈspektrəl]	a.	谱的
integrate	[ˈintigreit]	v.	积分
nest	[nest]	v.	嵌套
several-fold	[ˈsevərəl-fəuld]	ad.	翻几番
halve	[haːv]	v.	减半
MSL (mean sea level)	[miːn][siː][ˈlevəl]		平均海平面
MOS (model output statistics)	[ˈmɔdl][ˈautput][stəˈtistiks]		模式输出统计
regression	[riˈgreʃən]	n.	回归
predictor	[priˈdiktə]	n.	预报因子
thunderstorm	[ˈθʌndəstɔːm]	n.	雷暴
deficiency	[diˈfiʃənsi]	n.	缺乏
satellite-derived	[ˌsætəlait-diˈraivd]	a.	卫星导出的
impose	[imˈpəuz]	v.	强加
parameterize	[pəˈræmitəraiz]	v.	参数化
subgrid	[ˈsʌbgrid]	n.	次网格
unpredictable	[ˈʌnpriˈdiktəbl]	a.	不可预报的
precision	[priˈsiʒən]	n.	精密，正确
scattered	[ˈskætəd]	a.	分散的，稀疏的
subjective	[sʌbˈdʒektiv]	a.	主观的
frame-work	[freim-wəːk]	n.	框架
topographic	[ˌtɔpəˈgræfik]	a.	地形的
site-specific	[sait-spiˈsifik]	a.	定点的
gust	[gʌst]	n.	阵风
tornado	[tɔːˈneidəu]	n.	(陆)龙卷
freezing	[ˈfriːziŋ]	a.	冻结的
supplement	[ˈsʌplimənt]	v.	补充
array	[əˈrei]	n.	系列，数组

sounder	[ˈsaundə]	n. 探测器
acoustic	[əˈkuːstik]	a. 听觉的，有声的
Doppler (radar)	[ˈdɔplə][reidə]	n. 多普勒雷达
insect	[ˈinsekt]	n. 昆虫
refractive	[riˈfræktiv]	a. 折射的
index (indices)	[ˈindeks][ˈindisiːz]	n. 指数，索引
interpretation	[inˌtəːpriˈteiʃən]	n. 解释，说明
personnel	[ˌpəːsəˈnel]	n. 全体人员
software	[ˈsɔftwɛə]	n. 软件
prompt	[prɔmpt]	a. 迅速的，立即行动的，
		v. 激励，提醒
shear	[ʃiə]	n. 切变
downburst	[ˈdaunbəːst]	n. 下击暴流
predictability	[priˌdiktəˈbiliti]	n. 可预报性

Supplemental Reading

Rate of change

The local derivative $\frac{\partial}{\partial t}$ refers to the rate of change at a fixed point in rotating (x, y, z) space and the total time derivative d/dt refers to the rate of change following an air parcel as it moves along its three dimensional trajectory through the atmosphere. These so-called Eulerian and Lagrangian rates of change are related by the chain rule

$$\frac{d}{dt} = \frac{\partial}{\partial t} + u\frac{\partial}{\partial x} + v\frac{\partial}{\partial y} + w\frac{\partial}{\partial z}$$

which can be rewritten in the form

$$\frac{\partial}{\partial t} = \frac{d}{dt} - u\frac{\partial}{\partial x} - v\frac{\partial}{\partial y} - w\frac{\partial}{\partial z} \tag{1}$$

The terms involving velocities im Eq. (1), including the minus signs in front of them, are referred to as advection terms. At a fixed point in space the Eulerian and Lagrangian rates of change of a variable differ by virtue of the advection of air from upstream, which carries with it higher or lower values of. For a hypothetical conservative tracer, the Lagrangian rates of change is identically equal to zero, and the Eulerian rates of change is

$$\frac{\partial}{\partial t} = -u\frac{\partial}{\partial x} - v\frac{\partial}{\partial y} - w\frac{\partial}{\partial z}$$

(Adapted from Wallace and Hobbs, 2006)

4. The Greenhouse Effect

Text

Man is reversing millions of years of natural evolution by putting into the atmosphere carbon that had been sequestered over the ages as fossil fuels. Atmospheric concentrations of CO_2 are likely to double, and possibly triple, by 2100. Because no historical precedent exists, reasonable expectations about future climate must be based on scientific evidence, not geological records. After evaluating the available evidence, the National Academy of Sciences concluded that a doubling of atmospheric concentrations of CO_2 would warm the earth's average temperature 1.5—4.5℃.

The greenhouse effect (Fig. 1) of the atmosphere has never been doubted. More of the sun's radiation is visible light, which passes through the atmosphere largely undeterred. When the radiation strikes the earth, it warms the surface, which then radiates the heat as infrared radiation. However, atmospheric CO_2, water vapor, and some other gases absorb the infrared radiation rather than allow it to pass undeterredly through the atmosphere to space. Because the atmosphere traps the heat and warms the earth in a manner somewhat analogous to the glass panels of a greenhouse, this phenomenon is generally known as the "greenhouse effect". Without this effect, the earth would be 33℃ (60°F) colder that it is currently.

The extent to which CO_2 absorbs heat has been known for almost a century. Scientists show that doubling of atmospheric CO_2

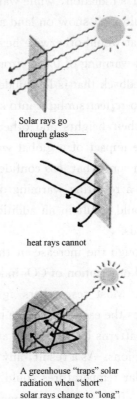

Solar rays go through glass

heat rays cannot

A greenhouse "traps" solar radiation when "short" solar rays change to "long" heat rays

Fig. 1 An illustration of the greenhouse effect

would raise the average temperature 1.2 ℃ if nothing else in the earth's climatic system changed. However, many parts of the climate will change, amplifying the direct impact of CO_2. Because these changes are not completely understood, the total warming is difficult to estimate. The current uncertainty surrounding the impact of CO_2 on average temperature is centered around these climatic "feedback", not the direct warming from CO_2.

The most important feedback will result from the warmer atmosphere's ability to retain more moisture. Because water vapor also absorbs infrared radiation, additional heating will result. Hanson et al. estimate that double CO_2 would increase the atmosphere's water vapor content 30%, heating the earth and additional 1.4℃.

Another important positive feedback concerns the impact of snow and ice cover on the earth's albedo, the extent to which it reflects sunlight. Ice and snow reflect most of the sun's radiation, while water and soil absorb it. An increase in surface temperatures would melt snow on land and floating ice and thereby allow the earth to absorb energy that would otherwise be reflected back into space. Hanson et al. estimate an additional warming of 0.4℃ from the albedo effect.

A feedback that is less understood is the impact of a global warming on clouds, which also reflect sunlight into space. The effects of clouds on the earth's albedo depend on their heights and other properties, as well as the extent of cloud cover. Thus, the impact of a global warming on clouds is somewhat uncertain. Nevertheless, with somewhat less confidence, Hanson et al. estimate a 2% reduction in cloud cover and a resulting warming of 0.5℃. They also estimate that increases in cloud height would result in an additional warming of 0.5℃, for a total impact of 1.0℃ from clouds.

Although the increase in the average temperature of the earth is a convenient shorthand description of CO_2 induced climate change, it masks important regional implications. Most researchers agree that polar temperatures would increase two to three times the earth's average increase. The world's climate depends largely on circulation patterns by which the atmosphere and the oceans transport heat from warm to cold regions. As a result, any significant change in the difference between equatorial and polar temperatures could dramatically affect climatic patterns. A particularly important effect of these changes will be shifts in annual and seasonal precipitation and evaporation, with some areas gaining and other losing. Furthermore, because hurricanes require an ocean temperature of 27℃ or warmer, a global warming could allow hurricanes to form at higher latitudes and during a greater part of the year. These changes could be important to coastal communities.

A more immediate concern is that the projected global warming could raise the sea as much as one meter in the next century by heating ocean water, which would then expand, and by causing mountain glaciers and parts of ice sheets in West Antarctica, East Antarctica, and Greenland to melt or slide into the oceans. Thus, the sea could reach heights unprecedented in the history of civilization until this effort. No one had attempted to forecast sea level rise in specific years or determine its importance to today's activities.

If this prediction comes true, the water will flood over 15 percent of Bangladesh's territory, about 12 to 15 percent of Egypt's cultivated land and significantly reduce the territory of many island countries. In the United States, thousands of square miles of land could be lost, particularly in low-lying areas such as the Mississippi Delta, where the land is also subsiding at approximately one meter per century. Storm damage, already estimated at over three billion dollars per year nationwide, could also increase the salinity of marshes, estuaries and aquifers, disrupting marine life and possibly threatening some drinking water supplies. The greenhouse gases are also damaging the ozone layer surrounding the earth which protects human beings from ultraviolet radiation. Research shows that cancer could increase by 3 percent for every 1 percent reduction in the ozone layer. Fortunately, the most adverse effects can be avoided if timely actions are taken.

Although the climatic change that could result from CO_2 emissions is poorly understood, there is complete agreement that CO_2 concentrations are increasing.

Approximately one-half the CO_2 released by combustion of fossil fuels has remained in the atmosphere. It is generally believed that most of the remaining CO_2 has dissolved into the oceans. Although tropical deforestation and cement production also result in CO_2 emission, their contributions have been and will continue to be much less important.

Controlling the greenhouse gases, especially the increasing discharge of carbon dioxide has already become a key problem. The global community must make efforts to reduce the greenhouse effect.

In the next few decades, however, CO_2 emissions are unlikely to be curtailed, either voluntarily or by regulation. The world's infrastructure is built around fossil fuels. The cost of using coal, gas and oil is low compared with nuclear and solar power, and this relative cost advantage is expected to continue. Therefore, a voluntary reduction in CO_2 emissions is unlikely.

The only government action that could successfully reduce CO_2 emissions would be to curtail the use of fossil fuels. Emission controls for CO_2 from power plants would at least quadruple the cost of electricity. For smaller users of fossil fuels, such as homes and motor vehicles, control is not even feasible. Other plans, such as sequestering carbon in massive tree plantings, are even less plausible.

Even if political leaders decide to take drastic actions to limit worldwide consumption of fossil fuels, it is probably already too late to prevent significant rises in global temperatures and sea level. A recent study investigated the impact of drastic energy policy changes on the expected timing of a greenhouse warming (Seidel and Keyes, 1983). The authors concluded that such policies could have important impact by 2100, but would not substantially delay the 2℃ warming expected by 2040. They estimated that a 300 percent tax on fossil fuels would delay the 2℃ warming by only five years, and that even a worldwide ban on coal, shale oil, and synthetic fuels would delay the warming by only twenty-five years, if implemented by 2000. Furthermore, such a ban would delay the rise in sea level expected through 2040 by only twelve years.

The political feasibility of instituting such a ban by 2000 is also doubtful, because only a worldwide agreement to curtail emission could be successful. Any individual nation that curtails its own emissions will delay the day when CO_2 concentrations double by a few years at most. Furthermore, because energy costs would increase for any nation that curtailed its emissions, that nation's industries would be placed at a competitive disadvantage compared with those of the rest of the world. Finally, political leaders would require proof that such a policy would be more beneficial than adapting to higher CO_2 levels. Such proof will probably remain impossible to provide for the foreseeable future.

The impact of increasing concentrations of greenhouse gases will almost certainly be an unprecedented global warming. Some people have suggested that this warming may be offset because the earth would otherwise be entering a cool period. However, a natural cooling would take place over tens of thousands of years and is thus unlikely to significantly offset the global warming in the next century. Even a drastic increase in volcanic activity would offset less than 10 percent of the projected rise in sea level.

To meet the challenge of a global warming, society will need accurate information concerning the likely effects of sea level rise. Unfortunately, communities, corporations, and individuals do not by themselves have sufficient resources or incentives to undertake the basic scientific research required to reduce existing uncertainties. This responsibility falls upon national governments throughout the world. Only their efforts can provide the information that decision makers will need.

New Words

evolution	[ˌiːvəˈluːʃən]	n.	进化,演化
sequester	[siˈkwestə]	v.	分离
triple	[ˈtripl]	v.	三倍于
precedent	[priˈsiːdənt]	n.	先例
geological	[dʒiəˈlɔdʒikəl]	a.	地质学的
academy	[əˈkædəmi]	n.	学院,研究所
National Academy of Sciences	[ˈnæʃənəl][əˈkædəmi][ˈsaiəns]		国家科学院
undeterred	[ˌʌndiˈtəːd]	a.	未受阻的
analogous	[əˈnæləgəs]	a.	相似的
panel	[ˈpænl]	n.	平板
amplify	[ˈæmplifai]	v.	放大
impact	[ˈimpækt]	n.	影响
feedback	[ˈfiːdbæk]	n.	反馈
retain	[riˈtein]	v.	保留,维持
shorthand	[ˈʃɔːthænd]	a.	简明的
mask	[maːsk]	v.	掩盖
implication	[ˌimpliˈkeiʃən]	n.	意义,含义
dramatically	[drəˈmætikəli]	ad.	鲜明地,惊人地
annual	[ˈænjuəl]	a.	(每)年的
coastal	[ˈkəustl]	a.	海岸的,沿岸的
immediate	[iˈmiːdjət]	a.	直接的,立即的
expand	[iksˈpænd]	v.	扩展,膨胀
glacier	[ˈglæsjə, ˈgleiʃə]	n.	冰川
Antarctica	[æntˈaːktikə]	n.	南极洲
Greenland	[ˈgriːnlənd]	n.	格陵兰
flood	[flʌd]	n.	洪水
Bangladesh	[ˌbaːŋgləˈdeʃ]	n.	孟加拉
Egypt	[ˈiːdʒipt]	n.	埃及
cultivate	[ˈkʌltiveit]	v.	耕作,种植

low-lying	[ləu-laiiŋ]	a.	低洼的
delta	['deltə]	n.	三角洲
Mississippi	[ˌmisi'sipi]	n.	密西西比河
subside	[səb'said]	v.	下沉
nationwide	['neiʃənwaid]	a.	全国范围的
salinity	[sə'liniti]	n.	含盐度
marsh	[mɑːʃ]	n.	沼泽,沼地
estuary	['estjuəri]	n.	港湾,河口湾
aquifer	['ækwifə]	n.	含水层
disrupt	[dis'rʌpt]	v.	破坏,瓦解
timely	['tainmli]	ad.	定时地,适时地
deforestation	[diˌfɔris'teiʃən]	n.	森林砍伐
cement	[si'ment]	n.	水泥
discharge	[dis'tʃɑːdʒ]	n.	释放,排放
curtail	[kəː'teil]	v.	减少
voluntarily	['vɔləntərili]	ad.	自愿地
regulation	[regju'leiʃən]	n.	管理,规定
infrastructure	['infrə'strʌktʃə]	n.	基础结构
voluntary	['vɔləntəri,(us)—teri]	a.	自愿的
quadruple	['kwɔdrupl]	v.	四倍于
plausible	['plɔːzəbl]	a.	似乎可能的
tax	[tæks]	n.	税
ban	[bɑːn]	n.	禁止
shale	[ʃeil]	n.	页岩
synthetic	[sin'θetik]	a.	合成的
implement	['implimənt]	v.	贯彻,完成
institute	['institjuːt]	v.	开始,制定,设立
competitive	[kəm'petitiv]	a.	竞争性的
disadvantage	[ˌdisəd'vɑːntidʒ]	n.	缺点,不利地位,劣势
lead	[liːd]	n.	带头
foreseeable	[fɔː'siːəbl]	a.	可预见的
offset	['ɔːfset]	v.	抵消,弥补
zone	[zəun]	v.	区划
undertake	[ˌʌndə'teik]	v.	从事于

5. The General Circulation of the Atmosphere

Text

The general circulation comprises the movements of the atmosphere on a worldwide scale. Since it is usually studied by means of data averaged over several days, so that minor, local or day-to-day irregularities are smoothed out, any model of the general circulation must be generalized, and cannot include very many short-lived features of importance for local weather. The general circulation is the overall pattern that must obviously affect local weather at sometime or another, directly or indirectly, and is in this sense the greatest single terrestrial cause of climate and weather.

Technically, the general circulation may be defined as the mean three-dimensional pattern of the meteorological elements, plus the "turbulence", the oscillation or perturbations of the mean pattern, provided by changing, day-to-day synoptic weather patterns. Its basic features may be described in terms of global, seasonal vector mean winds as a function of height, or they may be derived by applying the geostrophic wind relation to mean pressure-contour charts.

The three-dimensional aspects of the general circulation as actually observed must be particularly emphasized. For comparative purposes, it is convenient to separate the zonal (east-west) and meridional (north-south) components of the mean motion for the northern hemisphere. The mean meridional circulation is about a meter per second in the lower and middle latitudes throughout a substantial depth of the atmosphere; this is a much weaker circulation than the zonal one, but can nevertheless create or destroy momentum at the rate of 10 m per sec.

The simplest model that incorporates the main features of the observed mean meridional is given in the figures (omitted). The three kinds of cells are in the troposphere in each hemisphere; the Hadley cells in the tropics, the Ferrel cells in middle latitudes, and the weak subpolar cells beyond these. Angular momentum is injected into the Ferrel cells as indicated by the arrows, and is later carried downward by small convective eddies. Convection is most intense in low latitudes and thus for equilibrium to occur the Hadley cells must rotate faster than the Ferrel cells.

The model devised by Palmen takes into account the existence of jetstreams, which are the dominant features of the actual circulation. Observation shows that the Hadley cells, which are directly driven by heat, are the most important single elements of mean tropospheric circulation, but the Ferrel cells, driven by friction with the Hadley cells, prove to be more significant than Palmen envisaged.

An independent circulation is generated by heating and cooling in the stratosphere, down to about 10 hPa. Stratospheric winds reveal remarkable reversals in direction. A stratospheric monsoon occurs in the northern hemisphere: westerlies change to easterlies in April above 10 hPa, the reversal proceeding downward and southward from the polar regions, reaching 100 hPa in late May. Esterlies prevail above 100 hPa from May to August. In late August and early September, these easterlies revert back to westerlies.

Over North America in April, stratospheric polar easterlies are separated by the middle-latitude westerlies from the tropical easterlies of the lower atmosphere, which move northward in the month. By July, easterlies prevail down to at least 15 km in low latitudes, to 20 km in middle latitudes and to 15−17 km in polar latitudes. By September, the polar and tropical easterlies begin retreating to their minima, in November and December, respectively. In general, the picture at 10 hPa is of a slow, steady transition from summer easterlies to winter westerlies, but during January and February 1958 this simple pattern broke down in a very complex manner.

General models of the upper atmospheric circulation have been produced by Keliogg and Schilling (1951), Murgatroyd (1957) and Batten (1961). According to Betten's model, the major center of westerly winds is in the winter hemisphere, although these winds also cross the equator into the summer hemisphere. Easterlies occur in spring in the lower ionosphere, building down from the mesosphere as westerlies develop aloft; in turn, the westerlies then build down as easterlies develop aloft. Small easterly centers occur in the lower mesosphere in late winter and spring. The stratosphere winds above the Pacific equatorial region are extremely variable.

Important elements of the stratospheric circulation include the Berson westerlies and the Krakatoa easterlies. The former, first discovered at 50 to 60 hPa over central Africa, but now known to occur anywhere up to 10 hPa, form a continuous ribbon around the equator: the westerlies and easterlies alternate, one half-cycle being 12 to 15 months. The Krakatoa easterlies occur at 25 hPa: their existence was first inferred from the movement of volcanic dust after Krakatoa eruption. Radar wind observations now indicate that Krakatoa westerlies also occur.

The geographical importance of these stratospheric winds is that they make any simple, intuitive model of the atmospheric circulation untenable. In such a model, the rotation of the Earth from west to east is assumed to drag the lower part of the atmosphere with it, imparting a westerly motion to these layers. Thus slight variations in the momentum of the west-east rotating atmosphere would be interpreted at the Earth's surface as indicating winds apparently coming from different directions. A local excess of momentum in the atmosphere, causing the latter to move more rapidly from west to east than the Earth's surface, would be described as a west wind. A local deficit of momentum, causing the atmosphere to move less rapidly than the Earth's surface, would give rise to an east wind.

Recent observations indicate that the high atmosphere contains many circulation features that cannot be explained by a simple intuitive model. Thus, changes in the rotation of Sputnik 3 can be explained by the existence of a strong westerly wind, accompanying the Earth's rotation, but well above any region of possible frictional drag with its surface. Slow oscillations, representing a balance between inertia and Coriolis forces and static stability, have been measured by radarsonde theodolites at Crawley. They have period of 12 hours or so, and are due to disturbances with vertical and horizontal dimensions of 1 km and several 100 km, respectively. The complexity of data for the high atmosphere has made it necessary to split the circulations observed into mathematical components.

The circulation models discussed so far have been on diffusion. In Brewer's model, based on the diffusion of ozone and water vapor, air rises through the equatorial tropopause, which acts as a cold trap owing to its low temperature (around 80℃). The cold, dry air then moves horizontally, finally sinking in middle and high latitudes. According to Dobson's model, ozone-enriched air arriving via Brewer's meridional circulation is stored in the stratospheric polar-night jet, i. e., in the cold pool over the winter pole. From here, it gradually sinks into the lower stratosphere at temperate latitudes in late winter and spring.

In Spar's model, which is based on the diffusion of radioactive debris, the main exit for air from the stratosphere is through the gap in the tropopause, in which turbulent mixing takes place. More mixing takes place in the polar stratosphere (particularly in winter) than elsewhere, and much less mixing in the equatorial stratosphere than in the Brewer-Dobson model, which describes only one part of the whole circulation. The highest parts of the Brerrer-Dobson circulation reach 80,000 feet; above the atmosphere was envisaged by Brewer and Dobson as stagnant region is moist, and meridional transfer is affected by small-scale turbulent diffusion. The height of the transition from meridional-circulating to meridional-stagnant air varies both in time and in latitude.

In the Goldsmith-Brown model, rising air at the equator does not reach great heights, but turns poleward almost immediately above the tropopause. The meridional circulation is rapid just above the tropopause, the air taking slightly more than two months to reach temperate zones. The upper flow is much slower, so that air remains in the ozone-producing layers for about a year. From there, ozone-enriched air is fed slowly into the polar-night jet, where it becomes available for transfer into the middle-latitude lower stratosphere by the Dobson mechanism.

These circulation models for the high atmosphere are significant for weather at the Earth's surface because they demonstrate that air is definitely exchanged between troposphere and stratosphere. Surface weather systems, especially if described in terms of air masses and fronts, cannot be regarded as closed systems, although the error in so regarding them may not be serious in daily weather analysis. A further complication for climatological work is that the atmospheric and oceanic circulations must be considered a complexly integrated single system, if a complete study is to be made of the general circulation. If the stratospheric and oceanographic complications are ignored, the general circulation can be regarded as a consequence of part of the solar radiation received by the Earth being transformed into kinetic energy.

New Words

irregularity	[iˌreɡjuˈlæriti]	n.	不规则性
generalize	[ˈdʒenərəlaiz]	v.	概括,归纳
short-lived	[ʃɔːt-livid]	a.	短命的
vector	[ˈvektə]	n.	向量
aspect	[ˈæspekt]	n.	方面
zonal	[ˈzəunl]	a.	纬向的
component	[kəmˈpəunənt]	n.	分量
meridional	[məˈridiənl]	a.	经向的
subpolar	[ˈsʌbˌpəulə]	a.	副极地的
inject	[inˈdʒekt]	v.	注入,射入

equilibrium	[ˌiːkwiˈlibriəm]	n. 平衡
jetstream	[ˈdʒetstriːm]	n. 急流
envisage	[inˈvizidʒ,en-]	v. 观测,展望
reversal	[riˈvəːsəl]	n. 颠倒,反转
westerly	[ˈwestəli]	n. 西风
westerlies	[ˈwestəlis]	n. 西风带
easterly	[ˈiːstəli]	n. 东风
easterlies	[ˈiːstəlis]	n. 东风带
prevail	[priˈveil]	v. 盛行
revert	[riˈvəːt]	v. 颠倒,反转
retreat	[riˈtriːt]	v. 复原,恢复
equator	[iˈkweitə]	n. 赤道
ribbon	[ˈribən]	n. 带状物
infer	[inˈfəː]	v. 推导,推测
intuitive	[inˈtjuː(ː)itiv]	a. 直觉的,直观的
untenable	[ˈʌnˈtenəbl]	a. 守不住的,站不住脚的
drag	[dræg]	v. 拖曳
impart	[imˈpaːt]	v. 分给,传达
deficit	[ˈdefisit]	n. 亏损
Sputnik	[ˈspʌtnik]	n. (俄)苏联人造卫星
inertia	[iˈnəːʃiə]	n. 惯性
Coriolis	[ˌkɔːriˈəulis]	n. 科里奥利力
radarsonde	[ˈreidəˌsɔnd]	n. 雷达探空仪
theodolite	[θiˈɔdəlait]	n. 经纬仪
Grawley		n. 克鲁利(地方名)
complexity	[kəmˈpleksiti]	n. 复杂性
split	[split]	v. 分开
diffusion	[diˈfjuːʒən]	n. 扩散,漫射
trap	[træp]	n. 陷阱,冷阱(槽)
ozone-enriched	[ˈəuzəun-inˈritʃid]	a. 臭氧丰富的
radioactive	[ˈreidiəuˈæktive]	a. 放射性的
stagnant	[ˈstæɡnənt]	a. 滞停的
ozone-producing	[ˈəuzəun-prəˈdjuːsin]	a. 产生臭氧的
oceanographic	[ˌəuʃiənəuˈɡræfik]	a. 海洋学的
kinetic	[kaiˈnetik]	a. 运动的

Exercises (Reviewing and Thinking)

1. Answer the following questions according to the text by oral or writing

(1) What is the general circulation of atmosphere?

(2) What are the Hadley cells?

(3) What are the Ferrel cells?

(4) What are the subpolar cells?

2. Translate the following paragraph into Chinese

The simplest model that incorporates the main features of the observed mean meridional is given in the figure (Fig. 1). The three kinds of cells are in the troposphere in each hemisphere: the Hadley cells in the tropics, the Ferrel cells in middle latitudes, and the weak subpolar cells beyond these. Angular momentum is injected into the Ferrel cells as indicated by the arrows, and is later carried downward by small convective eddies. Convection is most intense in low latitudes and thus for equilibrium to occur the Hadley cells must rotate faster than the Ferrel cells.

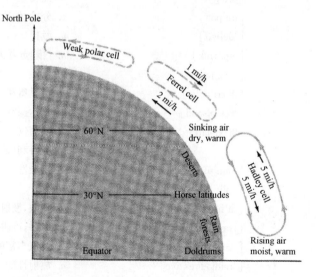

Fig. 1　The model of the mean meridional circulations of atmosphere

6. Severe Storms

Text

According to the Glossary of Meteorology, a storm is "any disturbed state of the atmosphere, especially as affecting the earth's surface and strongly implying destructive or otherwise unpleasant weather". In winter most storms are cyclonic in nature; they cover large areas, produce rain or snow, and often are cold and windy. These storms usually do more good than harm because the precipitation they yield waters farms and fills streams, lakes, and reservoirs.

Sometimes winter cyclones become intense as their central pressures fall. As a result the winds increase in strength, and if there is adequate atmospheric moisture, there can be flood-causing rains or heavy snow falls that close down entire cities. In some cases blizzards occur as strong, cold winds that carry the snow into large snowdrifts. In such storms, grazing animals on the open range can become isolated, run out of food or even freeze to death.

Among the most dangerous of winter storms are those that produce freezing rain. By coating streets, highways, and airport runways with ice, the storms cause a large number of motor vehicle accidents and pose serious hazards at airports.

Winter storms usually develop gradually, move slowly, and can be predicted with a reasonable degree of accuracy. On the other hand, severe storms of a convective nature develop so rapidly and are so small and short-lived that they are difficult to forecast. Violent thunderstorms, hailstorms, and tornadoes develop in tens of minutes and can strike with frightening suddenness. In minutes a hailstorm can destroy a field of wheat or a tornado can demolish a score of buildings, leaving injured and dead in its wake.

Another type of severe storms, the hurricane and its counterparts in other parts of the world, are relatively large, intense tropical cyclones. Although they do not contain the concentrated power of tornadoes hurricanes are more energetic larger and longer-lived, and therefore they can do more damage than tornadoes. Occasionally, individual violent storms can cause great losses. In some parts of the world extreme weather can take human tolls that nearly defy belief. For example, a tropical cyclone that struck Bangladesh in November 1970 took more than a quarter of million lives and rates as one of nature's most devastating calamities. The storm was a Southeast Asian version of the South Atlantic hurricane.

Commonly, thunderstorms extend to altitudes of about 10 km and often are topped by anvil clouds composed of the crystals blowing away from the main cloud regions.

Although thunderstorms take on many sizes, shapes and structures, they may be considered to fall into two broad categories: local or air mass thunderstorms and organized thunderstorms.

Local storms are fairly isolated and have a short lifetime——less than an hour or so. A great deal was learned by flying instrumented airplanes through thunderstorms and by observing them by means of radar, radiosonde and other ground-based instruments. A local thunderstorm is made up of one or more "cells" each of which follows a three-stage life cycle: cumulus stage, mature stage and dissipating stage.

Under the cloud, even before rain reaches the ground, there often is a downward rush of cool air that spreads rapidly outward away from the cloud. Depending on the strength of the upper-level winds and of the downdraft, outflowing surface winds can range from light breezes to strong blasts. The cool air usually advances ahead of the thunderstorm, and sometimes does appreciable damage to vegetation and buildings. The downdrafts and outflowing air and the associated wind shear represent a serious hazard to airplanes in the process of landing and taking off. Particularly strong downdrafts have been termed "downbursts" which are responsible for a number of crashes by commercial airlines.

Hurricanes

Tropical cyclones that occur over the Atlantic Ocean and the Gulf of Mexico and have maximum wind speeds greater than 32.6 m/s (73 mi/hr) are called hurricanes. Similar storms over the western North Pacific are called typhoons. The ones that develop over the Indian Ocean and affect India and Southeast Asia are called cyclones. Hurricanes off the Pacific coast of Mexico have been named cyclones by the natives of that country. Various other names are used in other parts of the world.

The paths followed by hurricanes are determined mostly by the prevailing wind currents in which the hurricanes are located. In their developing stages the storms are transported by easterly winds that prevail over the tropics. As the "steering" current changes, the hurricane path also changes. In some cases there are abrupt shifts in the direction of travel. As hurricanes move over land or over higher-latitude oceans, they become weaker. This occurs primarily because the energy input is reduced as the storm leaves the regions of warm ocean water. In addition, when a hurricane passes over a continent, the terrain exerts additional frictional force that act to reduce the speeds.

Hurricanes passing over land can be very destructive and lethal. When hurricane Camille swept into Mississippi in August 1969, it caused about 1.5 billion of damage. The torrential rains from hurricane Agnes in 1972 caused massive flooding over the eastern United States. Although it was not true in the case of Agnes, loss of life and destruction of property usually are a result of the storm surge, a wave of ocean water generated by the hurricane winds as the storm approaches a coast. A wall of water some 3 m or more in height can be produced. As it sweeps over low-lying land, it can cause devastating floods.

Hailstones

Hail usually is produced by large, long-lasting thunderstorm classified as multi-cell or supercell thunderstorms. The thunderstorms often are arranged in lines or bands that are called squall lines and occur most often in the spring and early summer months. These are the periods when atmospheric conditions are most favorable for the development of violent thunderstorms.

According to a highly regarded hailstorm model, first offered by two English scientists, hailstones began to form in the water-rich, rising air that enters the cloud along its leading edge. As they grow by colliding with supercooled drops, the hailstones fall at increasing speeds through the air and intercept increasing quantities of supercooled water. If the updraft speed exceeds the terminal velocity of the hailstones, they are carried upward. The upper-level winds move them forward across the upper part of the storm to a region of weaker updrafts. If the updrafts are not strong enough to hold up the hailstones, they fall toward the ground. After descending several kilometers, some of the stones fall into the strong updraft core and are carried upward again for a second passage through the supercooled cloud. A series of such trips can lead to large hailstones. An ice particle starting with a diameter of 1 mm can grow to 3 cm after exposure to the high liquid water contents in a strong updraft. In thunderstorms with strong, persistent updrafts, large hailstones can be grown without the recycling process.

Tornadoes

Thunderstorms producing large hail are also likely to produce tornadoes. Contrary to the opinions of many people, tornadoes are not particularly energetic. An average tornado contains much less energy than the thunderstorm that produces it, and is appreciably less energetic than a cyclone. But since the energy of a tornado is expended in a short period of time, its power-energy divided by time is relatively high. Because of their concentrated power, tornadoes are probably the most feared of all weather phenomena. They often strike suddenly, with little warning, and in a few minutes cause extensive damage to property, injuries and loss of lives. A tornado may have the appearance of a narrow funnel, cylinder, or rope extending from the base of a thunderstorm to the ground. The visible funnel consists mostly of water droplets formed by condensation in the funnel. Near the ground blowing dust, leaves and other debris identify the presence of a strong vortex. Tornadoes and weak visible vortex observed over water——most often in the tropics and subtropics——are called waterspouts.

Tornadoes are generally small, typically less than a few hundred meters in diameter, but some are larger than 1 km. With rare exceptions the winds are cyclonic, that is, they blow counterclockwise in the Northern Hemisphere. The funnels usually touch the ground for only a few minutes or so, but some have been reported to last for much more than an hour. The maximum wind speeds in tornadoes usually are between about 33 and 100 m/s (73 and 224 mi/hr), but occasionally they may exceed 120 m/s. The higher the intensity rating, the greater is the damage potential of a tornado. Most of the fatalities and property destruction are caused by a small number of large, long-lived tornadoes. During the decade ending in 1970, fewer than 2 percent of the largest tornadoes accounted for 85 percent of the fatalities.

Within a tornado there are still smaller, intense whirls that Fujita calls suction vortices. They have diameters of about 10 m. A small, short-lived tornado might have only one; a maxi-tornado would have many suction vortices. Fujita proposed that these small vortices account for occasional observations of virtually total destruction of one structure, while another one, 10 m away, is left unscathed.

The pressure in a tornado funnel is substantially lower than the surrounding atmospheric pressure. It has been estimated that in a very severe tornado the central pressure might he more than 100 hPa less than the pressure outside the funnel.

The pressure differences between the interior and exterior of a tornado can account in part for the damage to buildings whose doors and windows are closed tightly. When a tornado moves over such a structure, the outside pressure drops rapidly while the pressure inside falls more slowly. The resulting pressure differences augment those caused by the dynamical effects of the wind. When the pressure inside a building substantially exceeds the pressure on the outside, the result is a strong outward pressure force. Such a force can do extensive damage; in some instances it could pick up the roof and blow out the wall of a building.

Tornadoes——as is the case with hail——are most frequent in the late afternoon and early evening and occur in the spring and early summer months. It is during this period that the necessary meteorological conditions are most likely to be present.

Even in these years of space exploration, many tornadoes are detected visually by ordinary citizens who spot the funnels and notify local officials. Increasingly, radar is being used to detect and track tornado storms.

Conventional radars cannot specifically identify a tornado, and therefore it is necessary to use characteristics of the radar echoes to infer the presence of the tornadoes. Occasionally, the shape of an echo is a good indication of the presence of, or the impending formation of tornadoes. The most reliable such indicator is a hook-shaped echo extending from a thunderstorm, but most of the time the echo shape is of little use in ascertaining if there is a tornado. Thunderstorms that cause intense radar echoes that extend to great altitudes should be regarded as threats to produce hail and tornadoes.

A Doppler radar can observe the same quantities as can a conventional radar, but in addition it can measure the speed, toward or away from the radar of targets such as raindrops or other liquid or solid objects. Recent research has shown that Doppler radar can identify and locate tornadoes, sometimes as much as 20 minutes before the funnel reaches the ground.

New Words

glossary	['glɔsəri]	n. 词汇
reservoir	['rezəvwaː]	n. 水库
snowfall	['snəufɔːl]	n. 降雪
blizzard	['blizəd]	n. 雪暴

snowdrift	['snəudrift]	n.	雪堆
graze	[greiz]	v.	放牧
coat	[kəut]	v.	覆盖
violent	['vaiələnt]	a.	暴力的
hailstorm	['heilstɔ:m]	n.	雹暴
demolish	[di'mɔliʃ]	v.	破坏,推翻
wake	[weik]	n.	尾流
counterpart	['kautəpa:t]	n.	副本,对应的事物
toll	[təul]	n.	伤亡人数
defy	[di'fai]	v.	蔑视
rate	[reit]	v.	评价
devastating	['devəsteitiŋ]	a.	破坏性的
calamity	[kə'læmiti]	n.	灾难,灾害
Asian	['eiʃən]	a.	亚洲的
version	['və:ʃən]	n.	版本
top	[tɔp]	v.	封顶,盖住
anvil	['ænvil]	n.	砧
radiosonde	['reidiəusɔnd]	n.	无线电探空仪
enlarge	[in'la:dʒ]	v.	放大
downdraft	['daundra:ft]	n.	下曳气流
upper-level	['ʌpə-'lev(e)l]	n.	高层,高空
outflow	['autfləu]	v.	流出
blast	[bla:st]	n.	阵风
airline	['ɛəlain]	n.	航线,航空公司
typhoon	[tai'fu:n]	n.	台风
the Indian Ocean	['indjən]['əuʃən]	n.	印度洋
cyclone	['saikləun]	n.	气旋
native	['neitiv]	a.	本地人,土人
steer	[stiə]	v.	调整,导向
exert	[ig'zɛ:t]	v.	施加
destructive	[dis'trʌktiv]	a.	破坏性的
Camille	[kə'mi:l]	n.	开米丽(女子名,用以给台风命名)
sweep (swept, swept)	[swi:p]	v.	扫,扫荡
torrential	[tɔ'renʃəl]	a.	急流的,奔流的
inland	['inlənd]	n.	内陆,内地
multicell	['mʌltisel]	n.	多单体
supercell	['sju:pəsel]	n.	超级单体

initiate	[iˈniʃieit]	v. 开始,启动
water-rich	[ˈwɔːtə-ritʃ]	a. 含水丰富的
edge	[edʒ]	n. 边沿
exposure	[iksˈpəuʒə]	n. 暴露
recycle	[ˈriːˈsaikl]	v. 再循环
appreciably	[əˈpriːʃiəbli]	ad. 明显地
funnel	[ˈfʌnəl]	n. 漏斗
cylinder	[ˈsilində]	n. 圆筒
waterspout	[ˈwɔːtəspaut]	n. 水龙卷
cyclonic	[ˈsaikləunik]	a. 气旋性的
counterclockwise	[ˌkautəˈklɔkwaiz]	ad. 沿反时针方向
whirl	[(h)wəːl]	n. 涡旋,旋风
suction	[ˈsʌkʃən]	n. 虹吸,吸(水)
maxi-tornado	[ˈmæksi-tɔːˈneidəu]	n. 最大龙卷
unscathed	[ˈʌnˈskeiðd]	a. 未受损害的
exterior	[eksˈtiəriə]	n. 外部
tightly	[ˈtaitli]	ad. 紧紧地
augment	[ɔːgˈment]	v. 增大
device	[diˈvais]	n. 装置,设备
exploration	[ˌeksplɔːˈreiʃən]	n. 探索
intense	[inˈtens]	a. 强的
indication	[ˌindiˈkeiʃən]	n. 表示
impend	[imˈpend]	v. 迫近
indicator	[ˈindikeitə]	n. 指示器
hook-shaped	[huk-ʃeipt]	a. 钩状的

Supplemental Reading 6.1

Hurricane

Hurricane is a low pressure. A slight predominance of the inward-directed pressre gradient force continues to accelerate the air inward. The increased inflow also produces strong vertical velocities and increased condensation. Water vapor fuel for the convection is supplied at faster rates as the wind speed increases. The thermodynamic equation predicts warmer temperatures as a result of the condensation, and the warmer temperatures lead to still further decreases in surface pressure.

The intensification process does not continue indefinitely, however, because the balance of forces gradually changes For example, as the winds increase, so does the effect of friction, which tends to decrease the wind speed. The increased upward motion, besides the increasing convection, also increases the cooling due to expansion. Both the frictional and expansion cooling effects tend to weaken the circulation. At some point, these two weakening effects mat cancel or even slightly exceed, the the forces of intensification. At this point, the storm reaches a steady state in which the generation and dissipation effects are ballanced. Such a mature state is shown in Fig. 1.

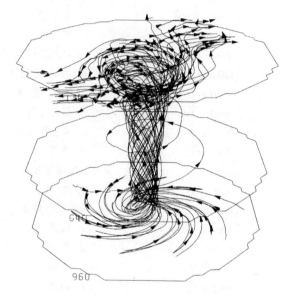

Fig. 1   Three dimensional trajectories in model hurricane (Adapted from Anthes, 1978)

Supplemental Reading 6. 2

Tornado

Tornadoes are similar to hurricanes in that strong winds blow counterclockwise around a center of very low pressure. In fact, the strongest surface wind on earth occur with tornadoes. Although the maximum tornadic velocities that have occurred are somewhat controversial (direct measurements being impossible), there is indirect evidence that winds hae exceeded 300mi/h in extreme case.

The diameter of the tornado is much smaller than that of the typical hurricane (Fig. 2) averaging about 1/4 kilometer. Thus, it is quite feasible to avoid an approaching tornado by driven ort even running at right angles to its path of motion. (Adapted from Anthes,1978)

Fig. 2　Picture of tornado

Supplemental Reading 6.3

Mesoscale Convective Systems (MCSs)

Mesoscale convective systems (MCSs) have regions of both convective and stratiform precipitation, and they develop mesoscale circulations as they mature. The upward motion takes the form of a deep-layer ascent drawn into the MCS in response to the latent heating and cooling in the convective region. The ascending layer overturns as it rises but overall retains a coherent layer structure. A middle level layer of inflow enters the stratiform region of the MCS from a direction determined by the large-scale flow and descends in response to diabatic cooling at middle-to-low levels. A middle level mesoscale convective vortex (MCV) develops in the stratiform region, prolongs the MCS, and may contribute to tropical cyclone development. The propagation of an MCS may have a discrete component but may further be influenced by waves and disturbances generated both in response to the MCS and external to the MCS. Waves of a larger scale may affect the propagation velocity by phase locking with the MCS in a cooperative mode. The horizontal scale of a MCS may be limited either by a balance between the formation rate of convective precipitation and dissipation of stratiform precipitation or by the Rossby radius of the MCV. (Adapted from Houze,2004)

Exercises

Translate the following paragraph into Chinese:

The discovery of the upper air wave as an even more important seat of energy than the associated low level cyclone, followed by the theoretical analyst of Rossby (based on Helmholtz's vorticity theorem) before World War II, and of such investigators as Charney, Eady, Fjortoft, Eliassen, Starr, Kuo, Lorenz, and Phillips, after the war, has finally led to the establishment of a fairly consistent picture of the large scale motions of the atmosphere, culminating in Phillips successful numerical solution for the main properties of the general circulation (1956).

7. Monsoon

Text

The term "monsoon" appears to have originated from the Arabic word mauims which means season. It is most often applied to the seasonal reversals of wind direction along the shores of the Indian Ocean, especially in the Arabian Sea, that blow from the southwest during one half of the year and from the northeast during the other. As monsoons have come to be better understood, the definition has been broadened to include almost all of the phenomena associated with the annual weather cycle within the tropical and subtropical continents of Asia, Australia, and Africa and the adjacent seas and oceans. It is within these regions that the most vigorous and dramatic cycles of weather events on the earth take place.

The dominant characteristic of the great monsoon system, the annual cycle itself, has led the inhabitants of the monsoon regions to divide their lives, customs, and economics into two distinct phases: the "wet" and the "dry". The wet, of course, refers to the rainy season, during which warm moist and very disturbed winds blow inland from the oceans. The dry refers to the other half of the year, when the wind reverses bringing cool and dry air from the hearts of the winter continents. In some locations, the cold and dry winter air flows across the equator toward the hot continents of the summer hemisphere. In this manner, the dry of the winter monsoon is tied to the wet of the summer monsoon, and vice versa.

In this article, we shall concentrate mainly on the annual cycle of the monsoon. However, it is incorrect to think of summer and winter phases of the monsoon as just prolonged periods of rain or drought, each of some months duration. There are also significant variations that exist on time scales ranging from days to weeks. Thus, while the monsoon appears to have a well-defined annual cycle, closer inspection shows that the monsoon varies substantially and that within the cycle a significant substructure exists that becomes evident as the intensity of the monsoon rains wax and wane through the wet season.

Short-term variations include the individual weather disturbances (i. e., a period of disturbed weather or storms lasting some days) that occur in rapid succession during the so-called active-monsoon periods. A prolonged period of one to several weeks marked by an absence of weather disturbances is called a break-monsoon, or more correctly, a break in the active monsoon. During an active phase the weather is unstable with frequent storms that produce the rain deluges traditionally associated with the monsoon. But, during a dormant or break phase of the monsoon, the weather is hot, clear, and dry. Monsoon breaks are drought period and, if prolonged, may cause considerable hardship and even famine in the monsoon lands.

A variable of the monsoon system of considerable importance is the timing of the commencement of the set. This, the so-called onset of the monsoon, is usually sudden with the weather changing abruptly from the pre-monsoon heat (similar to the torrid climate of the break-monsoon), to the weather disturbances, storms, and intense rainfall of an active period. For a farmer knowledge of when the onset will occur is critical as with it resides the key to the timing of the planting of his crops. The withdrawal of the monsoon (i. e., the cessation of rainfall over the continents) during the early autumn is a much more gradual transition than the onset.

Relationships between the Indian monsoon rainfall and the Southern Oscillation were established by Walker in the beginning of this century and versions of them have been used since for operational forecasting of monsoon rainfall. It is, therefore, useful to reexamine the relationship by using the Darwin sea level pressure for the period 1901—1981. Darwin pressure is chosen because its long-term record is considered to be more accurate and more complete than that for any other station in that region. Although Tahiti minus Darwin pressure is considered to be better index of the Southern Oscillation, Tahiti pressure is available only for the period 1935—1981, and for this period the correlation coefficient between the spring Tahiti pressure and Indian monsoon rainfall is only 0.01. The summer monsoon rainfall data used in this study are the area weighted average of the percentage departures for each of the 31 subdivisions of India, and is referred to as the whole Indian monsoon rainfall anomaly.

If one is going to predict monsoon rain, it is necessary to examine the Southern Oscillation before the monsoon season. However, it should be recalled that one of Walker's most important findings was that monsoon rainfall has very significant correlations with the subsequent global circulation.

The most remarkable of Walker's results was his discovery of the control that the Southern Oscillation seemingly exerted upon subsequent events and in particular of the fact that the index for the Southern Oscillation as a whole for the summer quarter June—August, had a correlation coefficient of 0.8 with the same index for the following winter-quarter, though only of -0.2 with the previous winter quarter. It is quite in keeping with this that the Indian monsoon rainfall has its connections with later rather than with earlier events. The Indian monsoon therefore stands out as an active, not a passive feature in world weather more efficient as a broadcasting tool than as an event to be forecast.

During the recent years we have gained a better understanding of large-scale atmospheric phenomena such as El Nino and the Southern Oscillation, the quasi-biennial oscillation and atmospheric blocking. This new knowledge provides a better synoptic and dynamical framework to examine the interannual and the longterm variability of monsoon. The El Nino-Southern Oscillation seems to be the single most important feature of the ocean-atmosphere systems. Its period is quite large (2—5 years) and therefore it can be of practical value for predicting fluctuations of a seasonal phenomenon like the monsoon. It is therefore necessary to document the main features of the global circulation, including the monsoon, during different phases of the Southern Oscillation. It should, however, be recognized that the fluctuations of the monsoon can also be one of the important factors affecting the Southern Oscillation.

The prospects for long-range forecasting of large-scale, seasonal mean monsoon rainfall appear to be good. There are significant correlations between large-scale seasonal mean Indian anomalies and low-frequency changes in the Southern Oscillation. There are also significant correlations between seasonal Indian rainfall anomalies and slowly varying boundary conditions of sea surface temperature and snow cover.

Tropical and monsoon flows are dominated by the thermally forced planetary scale Hadley and Walker type circulations for which the primary energy source is the latent heat of condensation. The larger-scale moisture convergence required for the release of the latent energy is organized by gradients of temperature at the earth's surface. Solar heating can produce thermal low-pressure areas over the land which can further deepen due to latent-heating if the dynamical circulation is favorable for moisture convergence. Therefore fluctuations of soil moisture can influence the intensity of the tropical heat sources over the land. Similarly, the tropical heat sources over the oceans can be influenced by the anomalies of sea surface temperature. It is therefore reasonable to expect that the changes in the large-scale tropical flows would be related to the changes in the slowly varying boundary conditions at the earth's surface. Since dynamical instabilities are not too strong in the tropics, it is also reasonable to hypothesize that the changes in the large-scale flows are dominated by the changes in the boundary conditions. These arguments collectively suggest that there is a physical basis for predictability of the large-scale, seasonally averaged monsoon flow and rainfall.

If the daily rainfall patterns related to the monsoon's high frequency, synoptic-scale disturbances were the consequence of dynamical instabilities of the large-scale flow, and if the changes of the large-scale flow itself were caused mainly by its interaction with such unstable? Forecasting beyond the limits of deterministic prediction would not be very good. Fortunately this does not appear to be the case. While it is indeed true that the rain producing disturbances form only when the structure of the large-scale flow (i. e., horizontal and vertical gradients of wind, temperature, and moisture) is favorable, the changes in the large-scale flow itself appear to be primarily related to planetary-scale boundary forcing manifested as tropical heat sources and to orographic barriers. This provides a physical basis as well as hope for long-range forecasting of monsoon rainfall. It is also of interest that during the monsoon season, even the biweekly and monthly anomalies have significant spatial coherence, which further suggests that the prospects for predicting biweekly and monthly anomalies are also quite good.

New Words

apply	[ə'plai]	vt.	运用,把…应用于
reversal	[ri'və:səl]	n.	逆转
adjacent	[ə'dʒeisənt]	a.	邻近的
vigorous	['vigərəs]	a.	强盛的
dramatic	[drə'mætik]	a.	引人注目的
inhabitant	[in'hæbitənt]	n.	居民
inland	['inlənd]	a.	内地的,内陆的
prolong	[prə'lɔŋ]	vt.	延长
duration	[djuə'reiʃən]	n.	持续时间
substantially	[səb'stænʃ(ə)li]	ad.	实质上,本质上
substructure	[sʌb'strʌktʃ(ə)]	n.	下部结构
wax	[wæks]	vi.	盈余
wane	[wein]	vi.	缺损,亏
individual	[ˌindi'vidjuəl]	a.	单一的,个别的
deluge	['delju:dʒ]	n.	泛滥,大洪水,倾盆大雨

dormant	[ˈdɔːmənt]	a.	休止的
famine	[ˈfæmin]	n.	饥荒
considerable	[kənˈsidərəbl]	a.	不可忽视的,重要的
timing	[ˈtaimiŋ]	n.	定时,时间选择
commencement	[kəˈmensmənt]	n.	开始
torrid	[ˈtɔrid]	a.	酷热的
reside	[riˈzaid]	vi.	留驻
cessation	[səˈseiʃən]	n.	停止
Southern Oscillation	[ˈsʌðən][ˌɔsiˈleiʃən]		南方涛动
version	[ˈvəːʃən]	n.	说法,不同看法
departure	[diˈpaːtʃə]	n.	偏差,出发
passive	[ˈpæsiv]	a.	消极的,被动的
quasi-biennial oscillation	[ˈkwaːzi-baiˈeniəl][ˌɔsiˈleiʃən]		准两年振荡
fluctuation	[ˌflʌktjuˈeiʃən]	n.	扰动,振荡
condensation	[kɔndenˈseiʃən]	n.	浓缩,凝聚(作用)
convergence	[kənˈvɜːdʒəns]	n.	聚合,汇合,辐合
gradient	[ˈgreidiənt]	n.	(温度,气压等的)变化率,梯度
anomaly	[əˈnɔməli]	n.	不规则,异常(现象)
hypothesize	[haiˈpɔθisaiz]	v.	假想
instability	[ˌinstəˈbiliti]	n.	不稳定性
orographic barrier	[ˌɔrəuˈgræfik][ˈbæriə]		地形分界线
spatial coherence	[ˈspeiʃəl][kəuˈhiərəns]		空间相干性

Supplemental Reading 7.1

The Computation of Equivalent Potential Temperature
ABSTRACT

A simplified procedure is described for computation of equivalent potential temperature which remains valid in situations such as in the tropics where a term which is omitted in the derivation of the conventional formula can lead to an error of several degrees absolute. The procedure involves new empirical formulas which are introduced for the saturated vapor pressure of water, the lifting condensation level temperature and the equivalent potential temperature. Error are estimated for each of these, and results are compared with those obtained by the similar, but more complicated procedures of Betts and Dugan (1973) and Simpson(1978).

8. A History of Numerical Weather Prediction

Text

In a short lecture it is not possible to do justice to all those who have made contributions to the development of numberical weather prediction (NWP), let alone those who have contributed in lesser ways. It should therefore be noted that this is "A" rather than "THE" history of NWP. The lecture covers the principal milestones in NWP from the "beginning" to the present day and is divided into three main sections.

1. The beginning

At the beginning of the twentieth century Bjerknes, the famous Norwegian meteorologist, was one of the first to consider forecasting the weather according to the dynamics of the atmosphere. Quite independently in 1911, L. F. Richardson who had already been interested in finite differences, began to dream of a forecast factory in which several thousand people would compute the weather. Once Richardson became aware of the Bjerknes work it had a considerable influence on his subsequent activities.

Due to interruptions during the war year, several revisions and problems with funding, it was 11 years before Richardson's (1922) now famous book on Weather Prediction by Numerical Process was published. He was a most interesting person and Ashford (1985) has written his biography. Richardson's book is remarkable on two accounts. He was one of the first to set out the fundamental equations of the dynamics and physics of the atmosphere in a systematic way and he produced a series of computing forms suitable for the numerical solution of those fundamental equations.

Richardson developed a set of partial differential equations for predicting pressure, temperature, horizontal and vertical velocity and moisture taking into account radiation, the effects of cloud and precipitation, eddy motion, friction, topography and exchanges of energy between the ground and the atmosphere which even included variations due to vegetation and state of ground. The consideration of these physical processes was most thorough and exhaustive. It is of interest that when the first realistic numerical integrations were carried out more than 20 years later they were purely dynamic; no physical processes being included. Indeed, it is only during the last 15 years that electronic computers have become fast enough to allow sophisticated treatment of physical processes to be gradually introduced into NWP models, although this was done somewhat earlier in GCM models where computational time was not critical.

In order to carry out the numerical integration of his prognostic equations, Richardson used latitude, longitude, height above mean sea level and time as the independent variables. He endeavoured to maintain a nearly square horizontal grid and discussed problems arising because of the polar cap and even considered the advantages of what is now known as a "sigma" coordinate system in the vertical. Although height was eventually used as the vertical coordinate, he chose his levels to be such as to allow the strata to contain approximately the same mass. His model contained 5 layers (6 levels at approximately 200 hPa intervals) and the horizontal grid was about 200 km although pressure, temperature and humidity alternated with momentum in the horizontal, making an effective grid length of 400 km. His consideration in depth of the choice of coordinate system is another example of how far ahead of his time Richardson was.

He quickly realized that in order to keep ahead of the weather he would need a very large number of people working on different part of the map. He therefore produced a most methodical system of 23 computing forms to allow the calculations to process speedily and accurately.

2. Further developments

The matter then lay almost dormant until the advent and availability of electronic computers at the end of World War II, although Rossby (1939) was developing the use of the vorticity equation and Sutcriffe (1917) the concept of development using thickness patterns; two ideas which were to play an important role in subsequent events.

The setting up of a Meteorological Research Group at the Institute of Advanced Study, Princeton, under Jule Charney, to consider the application of electronic computers to dynamic weather prediction was probably the most important single step in the history of the subject. The concept of a non-divergent level in midtroposphere where the absolute vorticity would be conserved, together with the filtering out of troublesome gravity waves from the vorticity equations by the geostrophic approximation, led to the computation of the first barotropic forecast. The computation time was almost exactly the same as the atmosphere took to achieve a similar result. It can be seen from the results of this 24 hour barotropic forecast that the northeastward movement, but not the deepening, of the 500 hPa vortex over the central U. S. A. was forecasted, as was the building up of pressure over and to the west of Oregon. This was most encouraging. It showed, in contrast to Richardson's earlier attempt, that NWP was now both possible and practical and led to the formation of many research groups throughout the world which have resulted in the excellent standard of today's products.

The development of the so-called filtered models progressed rapidly. Baroclinicity was introduced by two different approaches; the superimposition of several barotropic layers on the one hand and the adaptation of development theory on the other. Problems relating to the divergence of the geostrophic wind were overcome by the use of a stream function and empirical methods were developed for counteracting the spurious retrogression of the long waves. Methods of parameterizing physical processes such as surface friction, topography and non-adiabatic heating were gradually introduced. It is of interest to note that Cressman and Hubert (1957) showed that systematic errors caused by the divergence of the geostrophic wind could be almost eliminated by the use of a stream function and that it was essential to remove those errors before other errors in the treatment of baroclinic development and non-adiabatic effects could be isolated and resolved.

However, the speed of electronic computers was rapidly increasing and in the 1960s one could envisage being able to use a time step sufficiently short to ensure computational stability of the non-filtered (primitive) equation of motion and still compute the forecast in time to be of use. Charney (1955) had already demonstrated that if the initial wind and pressure fields were adjusted to eliminate the troublesome initial divergence, a satisfactory forecast could be produced by using the non-filtered equations. Eliassen (1956), Smagorinsky (1958), Hinkelmann (1959), Shuman (1962) and Bushby and Timpson (1967) all made significant contributions to the development of multi-level primitive equation models. Since then, as computers became faster, horizontal and vertical resolutions were increased and it has been possible to represent the physics in a more realistic manner.

The Oriental Influence was quite marked at that time as several eminent Japanese meteorologists moved to the U. S. A. and played a prominent role in the development of NWP. Arakawa's (1966) work on grid and convection schemes, Kasahara's (1965) on finite difference methods, Manabe's (1956) on the thermodynamic effects and Miyakoda's (1982) on extended range predictions are a few examples of the contributions that were made.

One of the first attempts to predict weather, as distinct from pressure patterns and vertical velocity fields was made by Bushby and Timpson (1967). A 10-level limited area primitive equation model was used with variables represented at 100 km intervals. The results were very encouraging. The 24 hour forecast of the fast moving wave depression showed good agreement with the verifying chart, as did the rates and areas of precipitation.

As computing speeds increased it became possible to increase the resolution of models and the area covered, and to introduce very sophisticated representation of physical processes. This had led to the present "state of the art".

3. Introduction into operational use

The first "real time" forecasts on a regular basis were produced by the Swedish Military Weather Service in 1954, in close cooperation with Rossby's group at the Meteorological Institute of Stockholm. They were in the nature of field trials and, after a second trial including an objective analysis scheme, it was decided that an NWP system was to be adopted as a basic operational feature from October 1, 1956. These first operational forecasts have been described by Bergthorsson, et al. (1955) and by Herrlin (1956). As shown in an example of 72-hour forecast, there was very encouraging agreement with the observed situations.

Much remains to be done. While the behaviour of temperate latitude cyclones is well represented in current models, the movement of tropical and subtropical disturbances especially into mid-latitudes, is not always well-handled. The development of non-hydrostatic mesoscale models pose new analysis and initialization problems, but will undoubtedly be necessary in order to forecast "weather" in detail. Finally, it must be remembered that there is a very strong correlation between the accuracy of forecasts and the sufficiency of the data-base from which they are made. All the research into improving NWP will be wasted if, at the end of the day, the data are unsatisfactory.

New Words

let alone	[let][əˈləun]	更不用说
milestone	[ˈmailstəun]	n. 里程碑
finite	[ˈfainait]	a. 有限的

subsequent	[ˈsʌbsikwənt]	a.	其后，接着
revision	[riˈviʒən]	n.	修正，订正
biography	[baiˈɔgrəfi]	n.	传记
partial	[ˈpɑːʃəl]	a.	不完全的
velocity	[viˈlɔsiti]	n.	速度
realistic	[riəˈlistik]	a.	逼真的
sophisticate	[səˈfistikeit]	vt.	使复杂
prognostic	[prɔgˈnɔstik]	a.	预报的
endeavour	[indevə]	n.	努力
strata（stratum 的复数）	[ˈstreitə][ˈstreitəm]		层
humidity	[hjuːˈmiditi]	n.	湿度
dormant	[ˈdɔːmənt]	a.	静止的
advent	[ˈædvənt]	n.	到来，出现
vorticity	[vɔːˈtisəti]	n.	涡度
conserve	[kənˈsəːv]	vt.	使守恒
barotropic	[ˌbærəˈtrɔpik]	a.	正压的
vortex	[ˈvɔːteks]	n.	涡度
baroclinicity	[ˌbærəkliˈnisiti]	n.	斜压性
superimposition	[ˈsjuːpərˌimpəˈziʃən]	n.	叠加
counteract	[ˌkauntəˈrækt]	vt.	抵消
spurious	[ˈspjuəriəs]	a.	假如
retrogression	[ˌretrəuˈgreʃən]	n.	退行
parameterize	[pəˈræmitəraiz]	v.	参数化
non-adiabatic	[ˈnɔn-ædiəˈbætik]	a.	非绝热的
envisage	[inˈvizidʒ,en-]	vt.	正视，面对，设想
primitive	[ˈprimitiv]	a.	原始的
prominent	[ˈprɔminənt]	a.	突出的
convection	[kənˈvekʃən]	n.	对流
thermodynamic	[ˈθəːməudaiˈnæmik]	a.	热力的
verify	[ˈverifai]	vt.	核实，验证
trial	[ˈtraiəl]	n.	试验
nonhydrostatic	[ˈnɔnˌhaidrəuˈstætik]	a.	非静力平衡
pose	[pəuz]	vt.	提出…问题

Supplemental Reading 8.1

For nearly a century meteorologists argued what constitutes "the perfect storm": "perfect" not in the sense of most catastrophic, but most typical of the cyclones generated by baroclinic instability in the real atmosphere. The case study featured in this section conforms in most respects to the classical Norwegian polar front cyclone model devised by H. Bjerknes and collaborators of the Vergen School during the 1920s for interpreting surface weather observations over the eastern North Atlantic and Europe. Chracteristic features of the archetypical Norwegian polar front cyclone, summarized in Fig. 1, include the strong cold front, the weaker occluded front, and the comma-shaped cloug shields in the development of extttratropical cyclones as envisioned in the Norwegian polar front cyclone model.

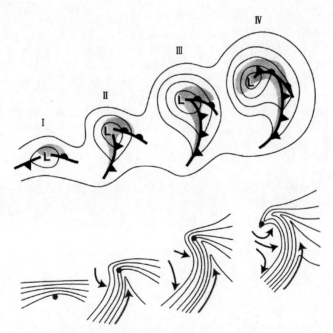

Fig. 1 Schematic showing four stage Panels Ⅰ, Ⅱ, Ⅲ and Ⅳ represent four sucessive stages in the life cycle of a cyclone (Adapted from Wallace and Hobbs, 2012)

Supplemental Reading 8.2

The life cycle of a Bjerknes and Solberg (1922) cyclone, hereafter the Norwegian cyclone model, begins with a small-amplitude disturbance on the polar front. This disturbance consists of a cyclonic circulation that advects cold air equatorward west of the cyclone center and warm air poleward east of the cyclone center, forming cold and warm fronts, respectively. Since the cold front is observed to rotate around the system faster than the warm front, the cold front eventually catches up to the warm front, forming an occluded front. Originally, Bjerknes and Solberg (1922) believed that this catch-up initially would occur away from the low center.

9. Ozone

Text

Ozone, which shields the biosphere from the lethal ultraviolet radiation from the sun, is probably the youngest member of the family gases that comprise the present natural atmosphere of the earth. More than 4.5 billion years ago, the sun, the earth, and the rest of the solar system condensed out of an interstellar cloud of gas and dust. As the heavier elements (such as silicon, iron, nickel) of the primordial solar nebula coalesced to form the solid earth, the lighter elements (carbon, nitrogen, oxygen) formed simple chemical compounds (methane, ammonia, water) with hydrogen, which was overwhelmingly the most abundant element in the nebula. These volatile compounds of hydrogen, carbon, hydrogen and oxygen were physically and chemically trapped within the solid earth. In the course of time, a new atmosphere formed around the earth whose ingredients were the trapped volatile compounds released or "outgassed" from the solid earth.

Ozone appeared in the atmosphere only after the appearance of molecular oxygen (O_2), the second most abundant constituent on the present atmosphere. It is important to note that O_2 was not present in the primitive atmosphere, because it was not outgassed from the earth's interior. According to one school of thought emergence of oxygen had to wait for about 2 billion years, until the advent of green plant photosynthesis. As the atmospheric levels of O_2 increased, so did the levels of ozone (O_3). It has been argued that the evolution of ozone controlled the migration of life from the safety of oceans onto the land. According to Berkner and Marshall's (1965) qualitative calculations, the "first critical level" of O_3 was reached when O_2 attained a level of $1E-2$ PAL. At this stage the O_3 column density ($1E18$ molecules per cm^2) was sufficient to restrict the ultraviolet zone of lethality to a thin layer close to the ocean surface. This greatly enhanced photosynthetic activity by permitting life to spread up to the vicinity of the ocean surface. The second critical level of O_3 was reached when O_2 was at 0.1 PAL. An ozone column density of about $5.6E18$ molecules per cm^2 at this stage was sufficient to absorb totally the lethal ultraviolet radiation and to enable life to migrate onto the land for the first time. Even though Berkner and Marshall's scenario has been questioned and is known about the geological history of the earth's atmosphere, this much is probably certain, that ozone is perhaps the youngest member of the natural atmosphere and that emergence of the ozone layer was instrumental in the emergence of higher forms of life on the land.

Although molecular nitrogen and oxygen were discovered as early as 1774, the recognition of ozone as a distinct chemical species came only after the advent of controlled electric energy. Apparently, the Dutch scientist van Marum was the first to observe that a peculiar odor resulted from passing an electrical discharge through oxygen. The substance causing the odor was identified in the laboratory by the Swiss chemist Schonbein, who noted that same strong odor occurred in oxygen generated from the electrolysis of water and in air when they were subjected to an electric discharge. He suggested that the substance might be a permanent feature of the atmosphere and thus deserved a name; in 1840 he proposed that it be called ozone (probably after the Greek word "ozien" ——to smell). The first chemical identification of ozone was probably due to Soret (1863) who stated that "a molecule of ozone is composed of 3 oxygen atoms".

Existence of ozone in the natural troposphere was chemically proven in 1858 with the aid of a test paper containing potassium iodide, which was developed by Schonbein. The first clear spectroscopic detection of ozone related to the atmosphere was made in 1880 by Chappuis who was able to measure eleven absorption bands attributable to this substance and coinciding with the telluric bands in the solar spectrum that were thought to be due to the selective absorption properties of the earth's atmosphere.

Because ozone was known to be produced efficiently by electric discharge, including lightning, the early belief was that ozone is distributed close to the earth's surface. Hartley (1881) was the first to point out that ozone is a normal constituent of the higher atmosphere and that it is in larger proportion there than near the earth's surface. The first satisfactory determination of the height of the absorbing medium was given by Lord Rayleigh in 1918, who based his estimate on observation of the solar spectrum at sunrise and sunset. He concluded that atmospheric ozone is largely confined to a layer between 40 to 60 km above sea level. Because of limitations of the experimental method, his estimate was only qualitatively correct. Later, greatly improved measurements by Gotz, Meetham and Dobson (1934) established the presently accepted bound to the stratospheric ozone layer, 15 km to 50 km in altitude with a maximum occurring near 25 km.

The first technique for measuring the total abundance of ozone in a vertical column was suggested by Fabry (1921). Dobson (1926) refined the method and developed instrumentation capable of making precise measurements; their technique is even today the standard for making ground-based measurements of total ozone. Basically, the Dobson instrument measures relative intensities at different wavelengths in the spectral range 300—740 nm after passage of the radiation through the atmosphere. From the data so obtained, the ozone column density is "unfolded". Ozone observations are also made with a broad band optical filter, mainly in the Soviet Union and Eastern Europe. However, those data are not always in agreement with Dobson data taken at the same place and do not seem to be as precise as the latter. For these reasons, only the Dobson instrument at Boulder, Colorado has been accepted as the international standard.

Techniques have also been developed to measure ozone in situ from aircraft and balloon platforms. These methods use a "wet" chemical system in which ozone reacts with a chemical (potassium iodide) in solution, or a gas phase system in which the reactions produce chemical luminescence of which the intensity can be measured. The most recent developments employ satellites that either measure the spectral radiance of backscattered solar ultraviolet (BUV) radiation or limb scan in which the absorption of solar radiation over an appropriate spectral segment is measured.

Soon after the discovery of the ozone layer in the upper atmosphere, the question of its formation arose. Warburg (1921) found that light of wavelength 253 nm is able to dissociate O_2, leading to the formation of two oxygen atoms. Actually, 253 nm is close to the limit of the O_2 dissociation spectrum. Only spectrum (195—260 nm) may penetrate deeply into the atmosphere and dissociate molecular oxygen in the O_2 bands and continuum that bear his name. Nevertheless, it was known by 1930 that solar radiation could penetrate to the altitudes of the ozone layer and rough estimates of the corresponding O_2 photolysis cross sections were available. This knowledge led Chapman (1930) to propose a series of processes, subsequently referred to as the Chapman mechanism, that could explain the formation of the ozone layer. Chapman's predictions were in excellent agreement with observations available at the time. According to Chapman reactions, molecules of oxygen gas absorb ultraviolet solar radiation, and as a result the molecules are dissociated to form oxygen atoms (O). Collisions of oxygen molecules and oxygen atoms and other particles lead to the formation of ozone molecules. They in turn absorb ultraviolet radiation and are dissociated to form oxygen molecules and oxygen atoms. In actuality, there are many other reactions involved in the production and destruction of ozone.

So far, only the photochemistry of ozone has been discussed. Actually, ozone can be transported vertically and horizontally by atmospheric motions from regions of formation to regions where it can be stored, for example, from above 30 km to lower altitudes where photochemical destruction mechanisms are weak. Ozone is transported horizontally in a rather complex manner that is dependent upon meteorological conditions and on geography, especially latitudes.

Because of the apparent effectiveness of catalytic agents in depleting ozone, scientists have been concerned about anthropogenic threats to the ozone layer. Johnstom (1971) suggested that nitrogen oxides deposited in the upper atmosphere by supersonic transport aircraft could degrade ozone abundance to unacceptable levels. In response to this threat, the U. S. government established and funded the Climatic Impact Assessment Program (CIAP) under the U. S. Department of Transportation. This effort was the large scale study of stratospheric ozone problems. Around this time, Molina (1974) suggested that free chlorine introduced into the upper atmosphere as a result of the photodissociation of man-made chlorofluoromethanes (CFMs) could also greatly enhance the catalytic destruction of ozone. CFMs were in wide use as refrigerants and aerosol propellants, and researchers become concerned that they might accumulate to high levels in the atmosphere.

Perhaps the most important aspect of the ozone layer, at least to the biosphere, is its absorption of solar ultraviolet radiation before it can reach the earth's surface. It has been known for a long time that ultraviolet radiation can be harmful to both animal and plant life. Its most striking effect is sunburn or erythema in humans. Repeated exposure to ultraviolet radiation may result in the occurrence of skin cancer, which can be broadly divided into (potentially lethal) melanoma and nonmelanoma. There is strong evidence that the occurrences of both types of cancer are latitude—dependent and that they correlate well with the mean ozone overburden at each latitude. There is a good connection between stratospheric ozone abundance and the effects on the biosphere.

Because ozone is a strong absorber of solar radiation, its presence in the upper atmosphere also results in local heating there. Indeed, it is the presence of ozone at high altitudes that leads to the formation of the stratosphere. The rate of energy deposition per unit mass in the ozone reaches a maximum at about 45 to 50 km altitude (near the stratopause). A minimum in the temperature (the tropopause) occurs at lower altitudes (8 to 16 km, depending upon latitude) so that the intervening layer is characterized by a strong temperature inversion and dynamically stable air.

New Words

shield	[ʃi:ld]	v. 遮盖，挡 n. 盾
lethal	[ˈli:θəl]	a. 致命的
interstellar	[ˈintə(:)ˈstelə]	a. 星际间的
silicon	[ˈsilikən]	n. 硅
nickel	[ˈnikl]	n. 镍
primordial	[praiˈmɔ:djəl]	a. 原始的，初级的
nebula	[ˈnebjulə]	n. 星云
methane	[ˈmeθein]	n. 甲烷
ammonia	[əˈməunjə]	n. 氨
overwhelmingly	[ˌəuvəˈwelmiŋli]	ad. 压倒一切地
volatile	[ˌvɔlətail]	a. 易挥发的，易发作的
ingredient	[inˈgri:diənt]	n. 成分
outgas	[ˌautˈgæs]	v. 喷出(气体)
molecular	[məuˈlekjulə]	a. 分子的
school	[sku:l]	n. 学派，流派
emergence	[iˈmə:dʒəns]	n. 出现
photosynthesis (photosyntheses)	[ˌfəutəuˈsinθəsis]	n. 光合作用
migration	[maiˈgreiʃən]	n. 迁徙
PAL=(Present Atmospheric Level)	[priˈzent] [ˌætməsˈferik] [ˈlev(ə)l]	现代大气水准
lethality	[liˈθæliti]	n. 致命性
vicinity	[viˈsiniti]	n. 附近
migrate	[maiˈgreit; ˈmaigreit]	v. 迁徙
scenario	[siˈnɑ:riəu]	n. 观点
instrumental	[ˌinstruˈmentl]	a. 有帮助的
recognition	[ˌrekəgˈniʃən]	n. 识别，认出，承认
species	[ˈspi:ʃiz]	n. 种类，式样，样品
Dutch	[ˈdʌtʃ]	a. 荷兰的
advent	[ˈædvənt]	n. 到来，出现
odor	[ˈəudə]	n. 气味
Swiss	[swis]	a. 瑞士的

electrolysis (electrolyses)	[ilek'trəlisis]	n. 电解
Greek	[gri:k]	a. 希腊的
potassium	[pə'tæsjəm]	n. 钾
iodide	['aiədaid]	n. 碘
spectroscopic	[ˌspektrə'skɔpik]	a. 光谱学的
attributable	[ə'tribjutəbl]	a. 可归因的
telluric	[te'ljuərik]	a. 地球的
selective	[si'lektiv]	a. 选择的,有选择性的
refine	[ri'fain]	v. 提炼
instrumentation	[ˌinstrumen'teiʃən]	n. 仪器装置
ground-based	[graund-beist]	a. 地基的
unfold	[ʌn'fəuld]	v. 展开
optical	['ɔptikəl]	a. 光学的
filter	['filtə]	n. 滤波器
Boulder	['bəuldə]	n. (美)博尔德市
Colorado	[ˌkɔlə'ra:dəu]	n. 科罗拉多州
luminescence	[ˌlu:mi'nesns]	n. 发光,辉光
employ	[im'plɔi]	v. 使用,利用
radiance	['reidiəns]	n. 辐射率
limb	[lim]	n. 边缘
segment	['segmənt]	n. 段节,部分
penetrate	['penitreit]	v. 穿透
continuum	[kən'tinjuəm]	n. 连续区,连续光谱
photolysis (photolyses)	[fəu'tɔlisis]	n. 光解
actuality	[ˌæktju'æliti]	n. 实际,现实
in actuality	[in][ˌæktju'æliti]	实际上
photochemistry	[ˌfəutəu'kemistri]	n. 光化学
catalytic	[ˌkætə'litik]	a. 催化的
agent	['eidʒənt]	n. (作用)剂,试剂,媒介(物),因素,代理人
deplete	[di'pli:t]	v. 减少,使…空
anthropogenic	[ˌænθrəpəu'dʒenik]	a. 人类起源和发展的
supersonic	['sju:pə'sɔnik]	a. 超音速的
degrade	[di'greid]	v. 降低
unacceptable	['ʌnək'septəbl]	a. 不可接受的
assessment	[ə'sesmənt]	n. 评价
CIAP (Climate Impact		

Assessment Programme)	['klaimit]['impækt]['prəugræm]	气候影响评价计划
Department of Transportation	[di'pɑːtmənt][ˌtrænspɔː'teiʃən]	交通部
chlorine	['klɔːriːn]	n. 氯
chlorofluoromethane (CFM)	['klɔːrəuˌfluərəu'miːθein]	n. 氯氟碳化物,氟利昂
refrigerant	[ri'fridʒərənt]	n. 制冷剂
propellant	[prə'pelənt]	n. 推进剂
melanoma	[ˌmelə'nəumə]	n. 黑素瘤
nonmelanoma	[nɔnˌmelə'nəumə]	n. 非黑素瘤
latitude-dependent	['lætitjuːd-di'pendənt]	a. 与纬度有关的
overburden	[ˌəuvə'bəːdn]	n. 过荷量
deposition	[ˌdepə'ziʃən; diː-]	n. 沉积
intervene	[ˌintə'viːn]	v. 插入,介入

Supplemental Reading 9.1

Moncrieff and Klinker (1997) inferred deep layer inflow into large mesoscale convective systems in TOGA COARE by a rather different approach. They simulated a TOGA COARE case within global model, which had a resolution of 80 km (T213). The model parameterized convection and cloud microphysics on the grid scale and resolved very large cloud clusters. Despite the coarse resolution, mesoscale convective systems formed in the model and exhibited realistic features such as the curved region of convective ascent and the mesoscale rear inflow under a broad anvil. One of the simulated features was a deep layer of inflow from ahead of the cloud system. Moncrieff and Klinker were concerned that the system might have been the result of aliasing smaller systems onto a large model-resolvable scale. However, systems of the size of that in Fig. 1 were indeed observed during TOGA COARE (e.g., see Chen et al., 1996). Although MCSs of this extreme size can occur and may sometimes be resolved by general circulation models (GCMs), there are likely many other situations in which smaller MCSs are aliased upscale in GCMs. Climate models likely miss MCSs altogether and fall back on parcel-based parameterizations of convection.

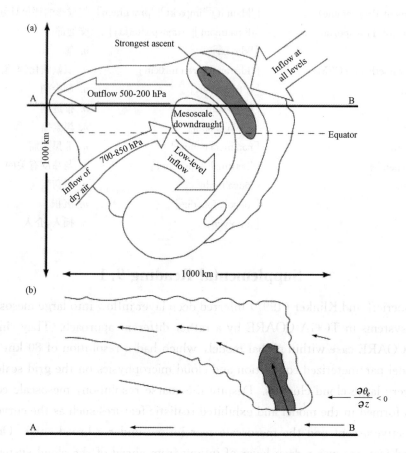

Fig. 1 Conceptual model of a supercluster, which is a large mesoscale convective system of the type that occurs over the western tropical Pacific. (a) Plan view and (b) zonal vertical cross section along line AB. Note the depth of the inflow layer at B. From Moncrieff and Klinker (1997)

10. Weather Hazards in Agriculture

Text

The growth of crops is not only dependent on weather conditions but crops are subject to a number of weather hazards until they are harvested. The major weather phenomena that constitute hazards to agriculture are frost, drought, hailstones and high winds. The nature of these phenomena, the hazards they pose to agriculture, and ways of managing them will now be described.

Frost

Frost is said to occur if the temperature of the air in contact with the ground (ground frost) or at screen level (air frost) is below 0℃. It is ground frost that is particularly important in agriculture. In weather forecasts, the term "ground frost" signifies a grass minimum-temperature below 0℃. There are two main genetic types of frost: radiation frost and advection or air mass frost. Radiation frost results from the rapid cooling of the air layer close to the ground following large terrestrial radiation losses on clear, calm nights. Advection or air mass frost occurs when an area is invaded by a cold air mass. Consequently, advection affects a large area when it occurs, whereas radiation frost tends to be spotty in occurrence.

Frosts are largely unknown in the tropics except in isolated mountainous areas. Frosts are, on the other hand, common in the temperate region and in the subtropical areas which suffer occasional incursions of cold air masses. Because the temperature of air masses cannot be controlled on a large scale, not much can be done to control or forestall the hazard posed by advection frost. Prevention of crop damage by radiation frost is, however, more feasible.

The purpose of frost prevention measures is essentially to break up the inversion that accompanies intense night-time radiation. This may be achieved in either of the following ways:

1. Heating in air by the use of the oil burners which are strategically located throughout the farm. The pall of smoke formed also reduces loss of heat by terrestrial radiation just like cloud cover does. The heat from the burners also creates convection

currents and brings about mixing of air in the inversion layer. The problem with oil burners is that the soot settles on crops and other things around. Also by daytime the smoke retards the warming of the ground surface. Heaters which give less smoke are now more commonly used.

2. Mixing or stirring the air by the use of giant fans operated by electric or gas driven motors. Light planes and helicopters are also used for the same purpose, particularly in mergence situation since these are rather costly.

The two methods above are used together and they work well provided the air is stable.

Other methods of frost protection include sprinkling crops with water, brushing, and the use of shelter belts (windbreaks). The object of sprinkling crops with water for frost protection is to reduce excessive cooling and increase the thermal conductivity of the ground. The latent heat released when water freezes ensures that the temperature of the plants does not fall below the freezing level as long as the change of state occurs. This method of frost protection, however, has some limitations. Heat gain during the day is limited just as heat loss during the night is retarded. Also, successive applications of water become progressively less effective and there are difficulties in determining when and how much water should be sprinkled on the crops.

The method of brushing involves putting a protective covering of craft paper over plants to reduce night time radiation loss. It is obviously a slow process that is labour demanding. Mulching can also inhibit frost. Windbreak may also be used to minimize the effects of frost, particularly advective frost. Windbreaks are, however, more usually used to prevent crop damage by high winds and to control rates of transpiration from farms. For any frost prevention method to be effective, advance warning of frost incidence is necessary. In this connection it is easier to forecast the occurrence of advective or air mass frost than that of radiation frost. This involves the analysis of characteristics of air masses and predictions of their movements and possible modifications by local features. Detailed forecasts of radiation frost are not feasible. Only warnings of potential danger can be made. This is because radiation frost is spotty in occurrence being determined largely by local features. Farms in areas of frost hazard should be equipped to monitor local temperature, humidity and wind so that individual farmers can determine the degree of frost hazard on their farms.

Drought

Drought constitutes a grave hazard to agriculture in both the temperate and tropical regions of the world. Although there are various definitions of the term "drought" there is general agreement that a drought can be said to occur whenever the supply of moisture from precipitation or stored in the soil is insufficient to fulfill the optimum water needs of plants. Four types of droughts can be identified. These are permanent, seasonal, contingent and invisible droughts.

Permanent drought is found in arid areas where in no season is precipitation enough to satisfy the water needs of plants. In such areas agriculture is impossible without irrigation through the whole crop season. Seasonal drought occurs in areas with well defined wet and dry seasons as in most parts of the tropics. The drought can be expected every year owing to seasonal changes in atmospheric circulation patterns. Agriculture is possible during the rainy season or with the use of irrigation during the dry season. Contingent and invisible droughts result from the fact that rainfall is irregular and variable. Contingent drought is characteristic of subhumid and humid areas and occurs when over a period of time the rain fails to fall. Contingent drought constitutes a serious hazard to agriculture because of its unpredictability. Invisible drought is different from the other types of drought in that it is less easily recognized. The other types of drought are evident by the wilting of crops or the lack of much vegetative growth. Invisible drought occurs any time the daily supply of moisture from the soil or falling precipitation fails to equal the daily water needs of plants. A slow drying of the soil results and crops fail to grow at their optimum rates. Crop yield is, therefore, less than the optimum. The need for supplemental irrigation to combat invisible drought is rather difficult to establish since the crops do not wilt. They just don't grow at their optimum rate which is partly determined by the quality of the seed and soil nutrient.

Since drought is a condition in which water need is in excess of available moisture, drought damage to growing crops can be prevented by decreasing the water needs of crops and/or increasing the water supply.

Crops with low water demands can be planted to reduce the water need. Similarly, drought-resistant crops and short-season crops should be planted instead of crops that require a lot of moisture or a long growing season with the attendant increase in probability of drought occurrence. Certain cultivation practices help to conserve soil moisture and these should be practiced in drought-prone areas.

The most effective method of combating drought is through provision of water artificially or by irrigation. Artificial stimulation of rainfall is at present an insignificant method of combating drought. On the other hand, irrigation is a widespread and common method of providing all or part of the water needs of crops. In an arid environment agriculture is possible only with irrigation. In semiarid and subhumid areas irrigation increases crop yield and the length of the growing season and this makes possible the cultivation of a greater variety of crops. In a humid region, irrigation helps to combat the effect of drought and increase crop yield.

Hailstones

Hailstones can physically damage young crops in the fields and so constitute a major hazard to agriculture wherever they occur frequently. Hailstones are hard pellets of ice of variable size and shape that fall from cumulonimbus cloud. Hail is, therefore, a solid form of precipitation. Hailstones occur both in the temperate region and in the tropics. Crop losses due to hailstones may be considerable. Hence, various measures are usually taken to minimize or prevent crop damage from hailstones. These have invariably involved the seeding of cloud with silver iodide released from artillery shells or aircraft. The aim of seeding is to create more small ice particles and more but smaller hailstones which are less damaging to crops.

Wind

Wind transports moisture and heat in the atmosphere and therefore has some effect on crop production. Wind also influences rates of evapotranspiration and directly exerts pressure on crops along its path.

Crop damage by winds may be minimized or prevented by the use of windbreaks (shelter belts). These are natural (e. g. trees, shrubs, or hedges) or artificial (e. g. walls, fences) barriers to wind flow to shelter animals or crops.

There is a reduction in both wind speed and rates of evaporation before the windbreak is reached. The decrease becomes sharper immediately downwind of the barrier and thereafter becomes less noticeable until wind speed and evaporation rates reach their prebarrier levels. Apart from influencing wind speed and rates of evaporation, air temperature and humidity, soil temperature and moisture are also altered over the area affected by the presence of a windbreak.

New Words

constitute	[ˈkɔnstitjuːt]	v.	构成,组成
frost	[frɔst, frɔːst]	n.	雾
drought	[draut]	n.	干旱
hailstone	[ˈheilstəun]	n.	(冰)雹块
pose	[pəuz]	v.	提出
screen	[skriːn]	n. v.	屏蔽,屏幕,百叶箱
signify	[ˈsignifai]	v.	表示
genetic	[dʒiˈnetik]	a.	发生的
radiation	[ˌreidiˈeiʃən]	n.	辐射
terrestrial	[tiˈrestriəl]	a.	地(球)的
invade	[inˈveid]	v.	侵入
spotty	[ˈspɔti]	a.	有斑点的
occurrence	[əˈkʌrəns]	n.	发生
isolated	[ˈaisəleitid]	a.	孤立的,隔绝的,偏僻的
subtropical	[ˈsʌbˈtrɔpikəl]	a.	副热带的
incursion	[inˈkəːʃən]	n.	侵入,袭击
forestall	[fɔːˈstɔːl]	v.	阻止,预防
prevention	[priˈvenʃən]	n.	预防
strategically	[strəˈtiːdʒikəli]	ad.	战略上
pall	[pɔːl]	n.	幕
soot	[sut]	n.	烟灰
retard	[riˈtɑːd]	v.	延缓,阻碍
heater	[ˈhiːtə]	n.	加热器
stir	[stəː]	v.	搅拌
sprinkle	[ˈspriŋkl]	v.	喷洒
brush	[brʌʃ]	v.	刷,擦
conductivity	[ˌkɔndʌkˈtiviti]	n.	传导率
successive	[səkˈsesiv]	a.	连续的
progressively	[prəˈgresivli]	ad.	渐渐地
craft paper	[krɑːft][ˈpeipə]	n.	牛皮纸
mulch	[mʌltʃ]	v.	覆盖
inhibit	[inˈhibit]	v.	防止,抑制
transpiration	[ˌtrænspiˈreiʃən]	n.	蒸腾
modification	[ˌmɔdifiˈkeiʃən]	n.	修改,改变
equip	[iˈkwip]	v.	装备

grave	[greiv]	a. 严重的
optimum	['ɔptiməm]	n. 最佳的
contingent	[kən'tindʒənt]	a. 级联的,列联的
arid	['ærid]	a. 干旱的
irrigation	[ˌiri'geiʃən]	n. 灌溉
subhumid	['sʌb'hju:mid]	a. 半湿的
unpredictability	['ʌnpriˌdiktəbiliti]	n. 不可预报性
wilt	[wilt]	v. (使)枯萎
supplemental	[ˌsʌpli'mentl]	a. 辅助的
combat	['kɔmbət]	v. 反对,与…斗争
nutrient	['nju:triənt]	n. 营养物
drought-resistant	[draut-ri'zistənt]	a. 耐旱的
attendant	[ə'tendənt]	a. 伴随(发生)的
cultivation	[ˌkʌlti'veiʃən]	n. 耕作,种植
drought-prone	[draut-prəun]	a. 易发生干旱的
artificially	[ˌa:ti'fiʃəli]	ad. 人为地
stimulation	[ˌstimju'leiʃen]	n. 激发,激励
semiarid	['semi'ærid]	a. 半干旱的
pellet	['pelit]	n. 粒
invariably	[in'veəriəb(ə)li]	ad. 不变地,一定,总是
seeding	['si:diŋ]	n. 播撒,催化
silver	['silvə]	n. 银
artillery	[a:'tiləri]	n. 大炮
shell	[ʃel]	n. 炮弹
suppress	[sə'pres]	v. 抑制
Kenya	['ki:njə;'kenjə]	n. 肯尼亚
evapotranspiration	[iˌvæpəuˌtrænspi'reiʃən]	n. 蒸散
minimize	['minimaiz]	v. 减至最小
hedge	[hedʒ]	n. 树篱
fence	[fens]	n. 篱笆
barrier	['bæriə]	n. 栅栏,障碍(物)
downwind	['daunwind]	a. 下风,顺风
prebarrier	[pri'bæriə]	a. 在栅栏前的

11. Polar-Orbiting Meteorological Satellites and Image Interpretation

Text

Weather satellites have been mapping our Earth and atmosphere over the past 30 years. The TIROS (Television and Infrared Observation Satellite) system; the predecessor of the following three generations of operational weather satellite systems, was marked by the launch of the first weather satellite, TIROS-1, in April 1960. Nine additional TIROS satellites launched in the period from 1960 to 1965 made the system a semioperational system. Each TIROS satellite carried a pair of miniature television cameras and in approximately half of the missions a scanning infrared radiometer and an earth radiation budget instrument were included with the instrument complement.

The commitment to provide routine daily worldwide observations without interruption in data was fulfilled in February 1966 by the introduction of the TIROS Operational System (TOS), which employed a pair of ESSA (Environmental Science Services Administration) Satellites. The second decade of meteorological satellites was introduced by the successful orbiting on January 23, 1970, of ITOS-1 (Improved TOS), the second-generational weather satellite, moving rapidly closer toward the objectives of the US National Operational Meteorological Satellite System. TIROS-1 provided, for the first time, day-and-night radiometric data in real time, as well as stored data for later playback. Global observation of the earth's cloud cover was provided every 12 hours with the single ITOS spacecraft as compared to every 24 hours with two of the ESSA satellites. The ITOS provided day-and-night imaging by means of Very High Resolution Radiometers (VHRR's) and medium resolution Scanning Radiometers (SR's). It included Vertical Temperature Profile Radiometers (VTPR's) for temperature sounding of the atmosphere and a Solar Proton Monitor for Measurements of proton and electron flux.

The third-generation operational polar-orbiting environmental satellite system, designated TIROS-N, completed development and was placed into operational service in 1978. This new series has a new complement of data-gathering instruments. One of these instruments, the Advanced Very High Resolution Radiometer (AVHRR), is a four/five-channel scanning radiometer, sensitive to visible, near infrared, and infrared radiation. Channel 1 (0.58—0.68 μm) measures the reflected solar radiation corresponding to the color yellow, orange and red. Very thin fog or stratus clouds as well as cirrus and cirrostratus are more transparent for yellow and orange light than visible light of shorter wave length. The primary use of Ch 1 is daytime cloud mapping. Channel 2 (0.725—1.10 μm) collects near infrared reflected sun radiation. This channel is used for land-water delineation. The reflectance of green leaves is much higher in this spectral range than in the visible light. Channel 3 (3.55—3.93 μm) senses infrared radiation of shorter wavelength. With the sun above the horizon this channel includes both reflected solar radiation and emitted terrestrial radiation. This channel has been found to be very useful in the detection of clouds and separation of cloud-filled pixels from clouds free. Surfaces with high reflectance in this band are observed as surfaces with higher brightness temperature than the actual temperature since a portion of the sun radiation is added to the near blackbody radiation emitted from the cloud. Large variation in the 3.7 μm band soon over cloud during daytime are due to different cloud reflectances. Channel 4 enables us to measure the temperature of a surface. A single measurement of radiant flux and a knowledge of the emissivity are sufficient to define the temperature. The spectral emissivity of water and snow is very close to 1.0. Soil and vegetation emissivities vary between 0.9 and 1.0. Ch 4 is the most common channel for image output from the satellites in the TIROS-N series. On later instruments in the series, a third IR channel (Ch 5) adds the capability for removing radiant contributions from water when determining surface temperatures. Prior to inclusion of this third channel, corrections for water vapor contributions was based on statistical means using climatological estimates of water vapor content.

The TIROS Operational Vertical Sounder (TOVS) onboard TIROS-N and follow-on satellites has been designed so that the acquired data permits calculation of (1) temperature profiles from the surface to 10 hPa; (2) water vapor content at three levels of the atmosphere, and (3) total ozone content. The TOVS system consists of three separate and independent instruments, the data from which may be combined for computation of atmospheric temperature profiles. The three instruments are (a) the High Resolution Infrared Radiation Sounder (HIRS), (b) the Stratospheric Sounding Unit (SSU), and (c) the Microwave Sounding Unit (MSU). The HIRS measures incident radiation in 20 spectral regions of the spectrum with a 15cm diameter optical system to gather emitted energy from earth's atmosphere. The instantaneous field of view (IFOV) of all the channels is stepped across the satellite track by use of a rotating mirror. This cross-track scan, combined with the satellite's motion in orbit, provides coverage of a major portion of the earth's surface. The energy received by the telescope is separated by a dichroic beam-splitter into longwave (above 6.4 μm) energy and shortwave (below 6.4 μm) energy, controlled by field stops and passed through bandpass filter, relaying optics to the detectors. In the shortwave path, a second dichroic beam-splitter transmits the visible channel to its detector. The SSU makes use of the pressure modulation technique to measure radiation emitted from carbon dioxide at the top of the earth's atmosphere. The primary objective of the instrument is to obtain data from which stratospheric (25—50 km) temperature profiles can be determined. This instrument is used in conjunction with the HIRS and MSU to determine temperature profiles from the surface to the 50 km level. The MSU, the first operational microwave temperature profile sounder flown on meteorological satellite, is a 4-channel Dicke radiometer making passive measurements in 4 regions of the 5.5 mm oxygen region. The antennas scan 47.4 degrees and either side of nadir is 11 steps. The beam width of the antennas is 7.5 deg (half power point), resulting in a ground resolution at the subpoint of 109 km, 6 times larger than that for the HIRS. This larger FOV and longer scan time is typical of microwave radiometers. The use of microwave frequencies for the remote sensing of the earth offers the very important advantages that clouds which do not contain large droplets or large ice crystals are virtually transparent to microwave radiation, so the MSU is an all-weather sounder.

The life of TIROS-N series may be further extended to the mid-1990s or beyond through improvement and enhancements of the spacecraft and sensors. How is the next generation of meteorological satellites? What time will the first satellites of the fourth generation be launched? No one knows because of both technological feasibility and budgetary trade-offs. The primary mission of the polar orbiting satellites is the global acquisition of vertical temperature profiles for use by the numerical forecasting models of the National Meteorological Center. The deficiencies in the present TOVS sounding capability are in the ability to obtain soundings in cloudy regions near the centers of storm systems and within active weather fronts. Sensor systems tuned to the microwave portion of the spectrum have shown that temperature profiles can be obtained in stormy regions except in areas of heaviest rain. So, the Advanced Microwave Sounding Unit (AMSU), a multichannel (9 to 20 Channels in the 10 to 90 GHz and even 187 GHz) microwave sounder, is under consideration for flight on the future polar-orbiting meteorological satellites.

Satellite images record cloud patterns from which one can deduce the three-dimensional structures of wind and pressure fields. This makes use of four factors. First, clouds are suggestive of the condition of stability in the atmosphere. Thus, cloud-free skies are associated with subsidence under anticyclonic conditions or the influence of very dry continental air whilst deep cloudiness over a wide area is indicative of instability. Secondly, the relative importance of vertical or horizontal motion in the troposphere can affect cloud patterns: small horizontal motion is associated with mottled, cellular cloud patterns whilst strong vertical motion is indicated by elongated cloud cells or cloud lines. Thirdly, strongly and complexly layered clouds are frequently generated along zones of contact and mixing between air streams possessing contrasting temperature and/or humidity characteristics. If clouds at different levels can be distinguished, variations in the direction and speed of the air flow at different levels in the troposphere can be detected. Finally, one should take note of the air masses and air streams caused by the underlying topography which creates the difference in cloud patterns over land and sea surfaces.

A useful first step in using satellite images is therefore to identify the different types of clouds. Clouds have been classified into three major types based on their forms: (1) cirrus or curly clouds; (2) stratus or layered clouds; (3) cumulus or lumpy clouds. Cloud types were conventionally identified visually from fixed positions on the ground. With the use of satellite image, the viewing is vertically, from above. Without stereoscopic viewing capability, it is extremely difficult to differentiate the cloud's altitude of occurrence. The lower spatial resolution of the satellite images also means that microscopic elements of the clouds cannot be depicted. However, Conover (1963) developed a genetic classification of clouds for use in the recognition of clouds from satellite images. His classification scheme was based on six characteristics: (1) cloud brightness relating to the depth of cloud, the nature of cloud constituents and the angle of illumination; (2) cloud texture relating to the degrees of smoothness of the clouds; (3) vertical structure relating to the structure and heights of clouds deduced from cloud shadows; (4) forms of cloud elements, relating to their degree of regularity; (5) patterns of cloud elements, relating to their degree of organization; (6) the size of the elements and patterns. Conover's scheme has been widely accepted.

Some progress has been made towards computer-assisted nephanalysis. Basically, this involves the use of computer algorithms to identify cloud types and to estimate cloud amount. Most workers made use of spectral features (i. e. cloud brightness) in the visible, water vapor and thermal infrared channels (singly or jointly) to classify cloud types with the computer. Textural features have also been used in combination with cloud brightness to analyse cloud type and cloud amount.

New Words

polar-orbiting	[ˈpəulə-ˈɔːbitiŋ]	a.	极轨的
predecessor	[ˈpriːdisesə]	n.	前辈
launch	[lɔːntʃ; laːntʃ]	v. n.	发射
semi-operational	[ˈsemi-ˌɔpəˈreiʃnl]	a.	半业务性的
miniature	[ˈminiətʃə]	n. a.	小型的
mission	[ˈmiʃən]	n.	使命，任务
radiometer	[ˌreidiˈɔmitə]	n.	辐射计
complement	[ˈkɔmpliment]	n.	定额装备
commitment	[kəˈmitmənt]	n.	许诺，委托
TOS (TIROS Operational Satellite)	[ˈtairəus] [ˌɔpəˈreiʃənl][ˈsætəlait]		泰罗斯业务卫星(即艾萨卫星)
ESSA (Environmental Science Services Administration)	[inˌvaiərənˈmentl][ˈsaiəns] [ˈsəːvis][ədminisˈtreiʃən]		(美)环境科学服务管理局
ITOS (Improved TIROS Operational Satellite)	[imˈpruːvid][ˈtairəus] [ˌɔpəˈreiʃənl][ˈsætəlait]		先进泰罗斯业务卫星
radiometric	[ˌreidiəuˈmetrik]	a.	辐射的
playback	[ˈpleibæk]	n.	回放
image	[ˈimidʒ]	v.	显像
VHRR (Very High Resolution Radiometer)	[ˈveri][hai][ˌrezəˈljuːʃən] [ˌreidiˈɔmitə]		甚高分辨率辐射计
SR (Scanning Radiometer)	[ˈskæniŋ][ˌreidiˈɔmitə]		扫描辐射计
VTPR (Vertical Temperature Profile Radiometer)	[ˈvəːtikəl][ˈtempritʃə] [ˈprəufail][ˌreidiˈɔmitə]		垂直温度廓线(探测)辐射计
proton	[ˈprəutɔn]	n.	质子
monitor	[ˈmɔnitə]	n.	监测器
AVHRR (Advanced Very High Resolution Radiometer)	[ədˈvaːnSt][ˌrezəˈljuːʃən] [ˌreidiˈɔmitə]		先进甚高分辨率辐射计
sensitive	[ˈsensitiv]	a.	敏感的
transparent	[trænsˈpɛərənt]	a.	透明的

delineation	[di͵lini'eiʃən]	n.	线条,示意图,描述
reflectance	[ri'flektəns]	n.	反射率
horizon	[hə'raizn]	n.	地平线
pixel	['piksəl]	n.	像素
brightness	['braitnis]	n.	亮度
vegetation	[͵vedʒi'teiʃən]	n.	植被,植物(总称)
remove	[ri'mu:v]	v.	去除
statistical	[stə'tistikəl]	a.	统计(学)的
climatological	[͵klaiməTə'lɔdʒikəl]	a.	气候(学)的
TOVS (TIROS Operational Vertical Sounder)	['tairəus] [͵ɔpə'reiʃənl] ['və:tikəl] ['saundə]		泰罗斯业务垂直探测器
onboard	['ɔn'bɔ:d]	prep.	在(船/舱/机/卫星上)
follow-on	['fɔləu-ɔn]	a.	后续的
HRIS (High Resolution Infrared Sounder)	[hai] [͵rezə'lju:ʃən] ['infrə'red] ['saundə]		高分辨红外辐射探测器
SSU (Stratospheric Sounding Unit)	[͵strætəu'sferik] ['saundiŋ] ['ju:nit]		平流层探测装置
MSU (Microwave Sounding Unit)	['maikrəuweiv] ['saundiŋ] ['ju:nit]		微波探测装置
incident	['insidənt]	a.	入射
instantaneous	[͵instən'teinjəs]	a.	瞬时的
track	[træk]	n.	轨迹
cross-track	[krɔs-træk]	a.	横跨轨迹的,与轨迹交叉的
telescope	['teliskəup]	n.	望远镜
dichroic	[dai'krəuik]	a.	双色的
beam-splitter	[bi:m-'splitə]	n.	波束分裂器
bandpass	['bændpa:s]	n.	带通
optics	['ɔptiks]	n.	光学,光学系统
detector	[di'tektə]	n.	检测器
antenna	[æn'tenə]	n.	天线
nadir	['neidiə; 'neidə]	n.	天底
width	[widθ]	n.	宽度
subpoint	['sʌb͵pɔint]	n.	下点,投影点
remote	[ri'məut]	a.	遥远的
sensing	['sensiŋ]	n.	传感
budgetary	['bʌdʒitəri]	a.	预算上的
trade-off	[treid-ɔ:f]	n.	折衷

acquisition	[ˌækwɪˈzɪʃən]	n.	获得
tune	[tjuːn]	n.	调谐
stormy	[ˈstɔːmi]	a.	多风暴的
AMSU (Advanced Microwave Sounding Unit)	[ədˈvaːnst][ˈmaikrəuweiv] [ˈsaundiŋ][ˈjuːnit]		先进微波探测装置
deduce	[dɪˈdjuːs]	v.	推演,推断
dimensional	[dɪˈmenʃənəl]	a.	维的,度的
factor	[ˈfæktə]	n.	因子
suggestive	[səˈdʒestɪv]	a.	有启发的,可供参考的
cloudiness	[ˈklaudinis]	n.	云量
indicative	[ɪnˈdɪkətɪv]	a.	表示特征
mottle	[ˈmɔtl]	v.	(使)带有斑点
cellular	[ˈseljulə]	a.	细胞的,单元的
elongated	[ˈiːlɔŋgeitid]	a.	拉长了的
contrast	[ˈkɔntræst]	v. n.	相对立,反差
distinguish	[disˈtiŋgwiʃ]	v.	区分
curly	[ˈkəːli]	a.	卷曲的,有卷毛的
lumpy	[ˈlʌmpi]	a.	钵状的
conventionally	[kənˈvenʃənli]	ad.	常规上
stereoscopic	[ˌsteriəˈskɔpik]	a.	立体的
differentiate	[ˌdifəˈrenʃieit]	v.	区分
spatial	[ˈspeiʃəl]	a.	空间的
microscopic	[maikrəˈskɔpik]	a.	微观的
depict	[dɪˈpɪkt]	v.	描绘,描述
scheme	[skiːm]	n.	方案
illumination	[ɪˌljuːmɪˈneɪʃən]	n.	照明,照射
texture	[ˈtekstʃə]	n.	纹理
smoothness	[ˈsmuːðnɪs]	n.	平滑度
regularity	[ˌregjuˈlæriti]	n.	规则性
computer-assisted	[kəmˈpjuːtə-əˈsistid]	a.	计算机辅助的
nephanalysis (nephanalyses)	[nefəˈnælisis]	n.	云分析
algorithm	[ˈælgəriðəm]	n.	算法

12. Micrometeorology

Text

1 Scope of Micrometeorology

Atmospheric motions are characterized by a variety of scales ranging from the order of a millimeter to as large as the circumference of the earth in the horizontal direction and the entire depth of the atmosphere in the vertical direction. The corresponding time scales range from a tiny fraction of a second to several months or years. These scales of motions are generally classified into three broad categories, namely, micro-, meso-, and macroscales. Sometimes, terms such as local, regional and global are used to characterize the atmospheric scales and the phenomena associated with them.

Micrometeorology is a branch of meteorology which deals with the atmospheric phenomena and processes at the lower end of the spectrum of atmospheric scales, which are variously characterized as microscale, small-scale, or local-scale processes. The scope of micrometeorology is further limited to only those phenomena which originate in and are dominated by the shallow layer of frictional influence adjoining the earth's surface, commonly known as the atmospheric boundary layer (ABL) or the planetary boundary level (PBL). Thus, some of the small-scale phenomena, such as convective clouds and tornadoes, are considered outside the scope of micrometeorology, because their dynamics is largely governed by mesoscale and microscale weather systems.

1.1 Atmospheric Boundary Layer

A boundary layer is defined as the layer of a fluid (liquid or gas) in the immediate vicinity of a material surface in which significant exchange of momentum, heat, or mass takes place between the surface and the fluid. Sharp variations in the properties of the flow, such as velocity, temperature and mass concentration, also occur in the boundary layer.

The atmospheric boundary layer is formed as a consequence of the interactions between the atmosphere and the underlying surface (land or water) over time scales of a few hours to about one day. Over larger periods the earth-atmosphere interactions may span the whole depth of the troposphere, typically 10 km, although the PBL still plays an important part in these interactions. The influence of surface friction, heating, etc., is quickly and efficiently transmitted to the entire PBL through the mechanism of turbulent transfer or mixing. Momentum, heat and mass can also be transferred downward through the PBL to the surface through the same mechanism.

The atmospheric PBL height varies over a wide range (several tens of meters to several kilometers) and depends on the rate of heating or cooling of the surface, strength of winds, the roughness and topographical characteristics of the surface, large-scale vertical motion, horizontal advections of heat, and moisture, and other factors. In the air pollution literature the PBL height is commonly referred to as the mixing depth, since it represents the depth of the layer through which pollutants released from the surface and in the PBL get eventually mixed. As a result, the PBL is dirtier than the free atmosphere above it. The contrast between the two is usually quite sharp over large cities and can be observed from an aircraft as it leaves or enters the PBL.

Following sunrise on a clear day, the continuous heating of the surface by the sun and the resulting thermal mixing in the PBL cause the PBL depth to increase steadily throughout the day and attain a maximum value of the order of 1 km (range $\approx 0.2-5$ km) in the late afternoon. Later in the evening and through the night, on the other hand, the radiative cooling of the ground surface results in the suppression or weakening of turbulent mixing and consequently in the shrinking of the PBL depth to a typical value of the order of only 100 m (range$\approx 20-500$ m). Thus the PBL depth waxes and wanes in response to the diurnal heating and cooling cycle. The winds, temperatures, and other properties of the PBL may also be expected to exhibit strong diurnal variations.

1.2 The Surface Layer

Some investigators limit the scope of micrometeorology to only the so-called atmospheric surface layer, which comprises the lowest one-tenth or so of the PBL and in which the earth's rotational or Coriolis effects can be ignored. Such a restriction may not be desirable, because the surface layer is an integral part of and is much influenced by the PBL as a whole, and the top of this layer, is not physically as well defined as the top of the PBL. The latter represents the fairly sharp boundary between the irregular and almost chaotic (turbulent) motions in the PBL and the considerably smooth and streamlined (nonturbulent) flow in the free atmosphere above. The PBL top can be easily detected by ground-based remote-sensing devices, such as acoustic sounder, lidar, etc., and can also be inferred from temperature, humidity and wind soundings.

However, the surface layer is more readily amenable to observation from the surface, as well as from micrometeorological masts and towers. It is also the layer in which most human beings, animals, vegetation live and variations in meteorological variables with height occur within the surface layer and consequently, the most significant exchanges of momentum, heat and mass also occur in this layer. Therefore, it is not surprising that the surface layer has received far greater attention from micrometeorologists and microclimatologists than has the outer part of the PBL.

1.3 Turbulence

Turbulence refers to the apparently chaotic nature of many flows, which is manifested in the form of irregular, almost random fluctuations in velocity, temperature, and scalar concentrations around their mean values in time and space. The motions in the atmospheric boundary layer are almost always turbulent. In the surface layer turbulence is more or less continuous, while it may be intermittent and patchy in the upper part of the PBL and is sometimes mixed with internal gravity waves.

Near the surface, atmospheric turbulence manifests itself through the flutter of leaves of trees and blades of grass, swaying of branches of trees and plants, irregular movements of smoke and dust particles, generation of ripples and waves on water surfaces, and variety of other visible phenomena. In the upper part of the PBL, turbulence is manifested by irregular motions of kites and balloons, spreading of smoke and other visible pollutants as they exit tall stacks or chimneys, and fluctuation in the temperature and refractive index encountered in the transmission of sound, light and radio waves.

2 Micrometeorology versus Microclimatology

The difference between micrometeorology and microclimatology is primarily in the time of averaging the variables, such as air velocity, temperature, humidity, etc. While the micrometeorologist is primarily interested in fluctuations, as well as in the short-term (of the order of an hour or less) averages of meteorological variables in the PBL or the surface layer, the microclimatologist mainly deals with the long-term (climatological) averages of the same variables. The latter is also interested in diurnal and seasonal variations, as well as in very long-term trends in meteorological parameters.

Despite the above-mentioned differences, micrometeorology and microclimatology have much in common, because they both deal with similar atmospheric processes occurring near the surface. Their interrelationship is further emphasized by the fact that long-term averages dealt with in the latter can, in principle, be obtained by the integration in time of the short-time averaged micrometeorological variables. It is not surprising therefore to find some of the fundamentals of microclimatology described in books on micrometeorology. Likewise, microclimatological information is found in and serves useful purpose in texts on micrometeorology.

3 Importance and Applications of Micrometeorology

Although the atmospheric boundary layer comprises only a tiny fraction of the atmosphere, the small-scale processes occurring within the PBL are useful to various human activities and are important for the well-being and even survival of life on earth. This is not merely because the air near the ground provides the necessary oxygen to human beings and animals, but also because this air is always in turbulent motion, which causes efficient mixing of pollutants and exchanges of heat, water vapor, etc. , with the surface.

Turbulence is responsible for the efficient mixing and exchanges of mass, heat, and momentum throughout the PBL. Through the efficient transfer of heat and moisture, the boundary layer turbulence moderates the microclimate near the ground and makes it habitable for animals, organisms and plants. The atmosphere receives virtually all of its water vapor through turbulent exchanges near the surface. Evaporation from land and water surfaces is not only important in the surface water budget and the hydrological cycle, but the latent heat of evaporation is also an important component of the surface energy budget. This water vapor, when condensed on tiny dust particles and other aerosols (cloud condensation nuclei), leads to the formation of fog, haze and clouds in the atmosphere.

Turbulence exchange processes in the PBL have profound effects on the evolution of local weather. Boundary layer friction is primarily responsible for the low-level convergence of flow in the regions of lows and highs in surface pressure, respectively. The frictional convergence in a moist boundary layer is also responsible for the low-level convergence of moisture in low-pressure regions. The kinetic energy of the atmosphere is continuously dissipated by small-scale turbulence in the atmosphere.

Almost one-half of this loss on an annual basis occurs within the PBL, even though the PBL comprises only a tiny fraction (less than 2%) of the total kinetic energy of the atmosphere.

New Words

micrometeorology	[ˈmaikrəuˌmiːtiəˈrɔlədʒi]	n. 微气象学
scope	[skəup]	n. 范围
circumference	[səˈkʌmfərəns]	n. 周长
microscale	[ˈmaikrəuskeil]	n. 微尺度
adjoin	[əˈdʒɔin]	v. 邻接，靠近
ABL (atmospheric boundary layer)	[ˌætməsˈferik][ˈbaundəri][ˈleiə]	大气边界层
PBL (planetary boundary layer)	[ˈplænitri][ˈbaundəri][ˈleiə]	行星边界层
fluid	[ˈfluː(ː)id]	n. 流体
span	[spæn]	v. 跨
topographical	[ˌtɔpəˈgræfikəl]	a. 地形的
mixing	[ˈmiksiŋ]	v. 混合
suppression	[səˈpreʃən]	n. 抑制
shrink	[ʃriŋk]	v. 变小，退缩
wax	[wæks]	v. 变大，增加
wane	[wein]	v. 变小，退缩
investigator	[inˈvestigeitə]	n. 研究人员
comprise	[kəmˈpraiz]	v. 包含
restriction	[risˈtrikʃən]	n. 限制
chaotic	[keiˈɔtik]	a. 杂乱无章的
streamline	[ˈstriːmlain]	n. 流线
acoustic	[əˈkuːstik]	a. 声的
lidar	[ˈlaidə]	n. 光雷达
amenable	[əˈmiːnəbl]	a. 服从的，顺从的
micrometeorolgist	[ˈmaikrəuˌmiːtiəˈrɔlədʒist]	n. 微气象学家
microclimatologist	[ˈmaikrəuˈklaiməˈtɔlədʒist]	n. 微气候学家
manifest	[ˈmænifest]	v. 证明，显示
random	[ˈrændəm]	a. 随机的
scalar	[ˈskeilə]	a. n. 标量(的)
patchy	[ˈpætʃi]	a. 不整齐的，不调和的
flutter	[ˈflʌtə]	n. 动摇
blade	[bleid]	n. 叶片
sway	[swei]	v. 摇摆，倾斜

stack	[stæk]	n. 堆，一堆；v. 堆叠
chimney	[ˈtʃimni]	n. 烟囱
transmission	[trænzˈmiʃən]	n. 传输，发射，透射
versus	[ˈvəːsəs]	prep. 对
microclimatology	[ˈmaikrəuˌklaiməˈtɔlədʒi]	n. 微气候学
interrelationship	[ˈintəriˈleiʃənʃip]	n. 内部关系
fundamental	[ˌfʌndəˈmentl]	a. n. 基础，基本
well-being	[welˈbiːiŋ]	n. 福利
survival	[səˈvaivəl]	n. 幸存，存活
moderate	[ˈmɔdərit]	v. 调节，使适度
microclimate	[ˈmaikrəuklaimit]	n. 微气候
habitable	[ˈhæbitəbl]	a. 适于居住的
organism	[ˈɔːgənizəm]	n. 有机物
hydrological	[ˌhaidrəˈlɔdʒikəl]	a. 水文学的
haze	[heiz]	n. 霾

13. Approaches to Climatic Classification

Text

As a fundamental tool of science, classification has three interrelated objectives: to bring order to large quantities of information, to speed retrieval of information, and to facilitate communication. Classification of climate shares these objectives. It is concerned with organization of climatic data in such a way that both descriptive and analytical generalizations can be made, and it attempts to store information in an orderly manner for easy reference and communication, often in the form of maps. The value of a systematic arrangement of climates is determined largely by its intended use: a system that suits one purpose is not necessarily useful for another. For example, a classification based on critical temperature and moisture limits for growth of a certain plant or animal organism might serve the needs of a biological study, but it is not likely to be satisfactory for weather forecasting, which relies more on such factors as the general circulation, storm types and weather probabilities. Thus, in the design of a climatic classification we should begin by defining the purpose. Three broad approaches are equally feasible: (1) empirical, (2) genetic, (3) applied. Together they constitute a classification of classifications, but the features of all three may be incorporated in a single system.

Empirical classifications are based on the observable features of climate, which may be treated singly or in combination to establish criteria for climatic types. Temperature criteria, for example, might yield "hot", "warm" "cool" and "cold" climates, each of which can be defined in terms of strict mathematical limits. Adding precipitation and other elements to the criteria, the number of possible combinations rapidly multiplies, and soon the system becomes unwieldy. It is, therefore, necessary to select the criteria that are most significant in light of the intended purpose. Heat and moisture factors have dominated empirical classification, but all elements are inherently significant for one purpose or another.

Genetic classification attempts to organize climates according to their causes. Ideally, the criteria employed in the differentiation of climatic types should reflect their origins if climatology is to be explanatory as well as descriptive. In practice, however, explanations are often theoretical, incomplete, and difficult to quantify. Genetic classification also is subject to theoretical biases: a system based on causes tends to perpetuate faulty or over-generalized theories. The ancient Greeks recognized a relationship between latitude and temperature and devised a system of climate, or zones (torrid, temperate and frigid) that have persisted in writings to the present day in spite of evidence that net radiation does not vary solely with latitude and that other factors affect the world patterns of temperature. Besides latitude, features of the general circulation, including winds, air masses and storm types; terrain features such as elevation, slope and mountain barriers; and the distribution of land and water are other bases for genetic classification. A common genetic approach attempts to distinguish the relative continentality or maritimity (sometimes termed oceanity) of a climate. In practice, indices to express the influences of land or water surfaces have been determined from various empirical data, mainly temperature, precipitation, wind and air-mass frequency. The most widely accepted criterion is the mean annual range of temperature, which tends to be greater over the continents than over oceans. Since the annual temperature range is also a function of latitude, compensating adjustments are needed. The map (omitted) shows North American regions of relative continentality derived mainly from temperature ranges and latitude. Outside the middle latitudes it is difficult to isolate influences of land or sea in terms of temperature or other climatic variables. In the tropics annual temperature ranges usually are small even in continental interiors; at high latitudes complications are introduced by the polar night and ice cover. The concept of relative continentality-maritimity has its greatest taxonomic utility between about 30 and 60 deg. latitudes. Several formulas have been devised to correct for latitude as well as eliminate negative index values and allow for asymmetric annual marches of temperature. Early indices expressed continentality as a percentage, implying the existence of a shelly continental or wholly maritime climate, which is a genetic impossibility in the dynamic atmosphere.

Applied (also known as technical or functional) classifications of climate assist in the solution of specialized problems that involve one or more climatic factors. They define class limits in terms of the effects of climate on other phenomena. Outstanding among modern attempts at climatic classification are those that seek a systematic relationship between climatic factors and the world pattern of vegetation. Natural vegetation integrates certain effects of climate better than any instrument that has so far been designed, and it is thus an index of climatic conditions. By referring to the major plant associations, biologists have tried to determine the climatic factors that correspond with areal differences in vegetations. Numerous correlations between vegetation and heat or moisture factors have been discovered, permitting the use of temperature or moisture indices as criteria for climatic types. The resulting types and their regional boundaries approach reality in terms of the associated vegetation, while retaining a climatic basis. Commonly, classifications of this kind employ vegetation terms. Rain forest, desert, steppe and tundra are terms that have a climatic connotation. In each case there are climatic limits beyond which the characteristic plant association (or a specific indicator species) does not occur naturally. Whereas fluctuations in climate and in non-climatic influences create problems in delineations of static areal boundaries, they are of great significance in interpreting vegetation changes through time.

It should be evident that there are many possible classifications of climate, for classification is a product of human ingenuity rather than a natural phenomenon. A complete classification should provide a system of pyramiding categories, ranging from the innumerable microclimates of exceedingly small areas (and often restricted to a shallow layer at the earth's surface), through mesoclimates, to highly generalized macroclimates which is of a world scale. But the description of world climates is not easily accomplished as the summation of a great number of microclimates. Nor microclimates easily fitted into the pattern of major climatic regions. The higher categories of any classification system are necessarily generalizations; the lowest category must include individuals. Great difficulties attend the delimitation of an "individual climate", for climates vary as a continuum over the entire earth. The concepts of topoclimate, representing the climatic response to local topographic conditions, and ecoclimate, the climatic environment of a living organism, are useful approximations. In any event, an individual climate is the synthesis of all the climatic elements in a unique combination that results from interacting physical processes. Since the exchanges of energy and mass between the air and the earth's surface are basic climatic processes, both surface and atmospheric conditions are appropriate criteria for classification of climate. Large-scale generalization which take into account horizontal as well as vertical transport processes can be derived from a synthesis of atmospheric flow patterns. Sequences of synoptic maps, upper-air charts, and vertical cross sections afford a basis for classifying mean regional circulation types and their associated weather. This approach, known as synoptic climatology, has great utility in long-range weather forecasting and aids the study of both genesis and effects of climate.

In order to achieve objectivity in defining the categories of a system, it is useful to have quantitative measurements of the climatic elements. In the past the lack of adequate records with respect both to periods covered and to world-wide distribution has presented a serious obstacle. The more than 100,000 surface weather-observing stations of all type in the world are by no means evenly distributed, and many of them record only one or two climatic elements during short or irregular periods. Although 3,500 ships take meteorological observations that are of great value in weather forecasting, the transient nature of these stations limits the use of their records for climatological analyses. The small number of stationary weather ships provides only a token record of climate for huge expanses of water. Eventually, these problems should be overcome by satellite technology, which already provides observations of surface temperatures and offers the promise of a thorough charting of climates. But it still will be necessary to infer climatic data from the evidence of geology, glaciology, biology, archaeology, or early historical accounts in order to reconstruct climates of the distant past.

New Words

classification	[ˌklæsifiˈkeiʃən]	n.	分类(法)
interrelate	[ˌintə(:)riˈleit]	vt.	相互关联
objective	[əbˈdʒektiv]	a.	目标,目的
retrieval	[riˈtri:vəl]	n.	更正,修正
facilitate	[fəˈsiliteit]	vt.	使更容易
generalization	[ˌdʒenərəlaiˈzeiʃən]	n.	归纳
analytical	[ˌænəˈlitikəl]	a.	分解的,分析的
reference	[ˈrefrəns]	n.	参考,依据
intended	[inˈtendid]	a.	预期的
incorporated	[inˈkɔ:pəreitid]	a.	合并的,结合的
criterion (pl. criteria)	[kraiˈtiəriən]	n.	标准,准则
combination	[ˌkɔmbiˈneiʃən]	n.	组合
unwieldy	[ʌnˈwi:ldi]	a.	难使用的,不便利的,笨重的
inherently	[inˈhiərəntli]	ad.	内在地,固有地
ideally	[aiˈdiəli]	ad.	理想地,典型地
climatology	[ˌklaiməˈtɔlədʒi]	n.	气候学
quantify	[ˈkwɔntifai]	vt.	确定…的数量
perpetuate	[pə:ˈpetjueit]	vt.	继续,使永久存在

faulty	[ˈfɔːlti]	a.	有缺点的,不完善的
frigid	[ˈfridʒid]	a.	寒冷的
indices (index 的复数)	[ˈindisiːz]复:[ˈindeks]	n.	标准,指标
compensate	[ˈkɔmpənseit]	vt.	补偿
derive	[diˈraiv]	vt.	得到
taxonomic	[tæksəˈnɔmik]	a.	分类学的,分类的
eliminate	[iˈlimineit]	vt.	消除,消去
negative	[ˈnegətiv]	a.	负的
asymmetric	[æsiˈmetrik]	a.	不对称的
percentage	[pəˈsentidʒ]	n.	百分数
wholly	[ˈhəuli]	ad.	完全地
maritime	[ˈmæritaim]	a.	海的,海上的
vegetation	[ˌvedʒiˈteiʃən]	n.	植被
integrate	[ˈintigreit]	vt.	使…结合起来
steppe	[step]	n.	干草原
tundra	[ˈtʌndrə]	n.	苔原,冻原,寒漠
connotation	[ˌkɔnəuˈteiʃən]	n.	含义
delineation	[diˌliniˈeiʃən]	n.	描写
interpret	[inˈtəːprit]	vt.	说明,解释
evident	[ˈevidənt]	a.	明白的,明显的
ingenuity	[ˌindʒiˈnjuːiti]	n.	机智,独创性
pyramid	[ˈpirəmid]	n.	金字塔
exceedingly	[ikˈsiːdiŋli]	ad.	非常地
restrict	[risˈtrikt]	vt.	限制
restricted	[risˈtriktid]	a.	范围狭窄的
summation	[sʌˈmeiʃən]	n.	总结
delimitation	[diˌlimiˈteiʃən]	n.	区划,划界
individual	[ˌindiˈvidjuəl]	a.	单一的,个别的
topographic	[ˌtɔpəˈgræfik]	a.	地形(学上)的
synthesis	[ˈsinθisis]	n.	综合
genesis	[ˈdʒenisis]	n.	起源
synoptic	[siˈnɔptik]	a.	天气的,天气尺度的
sequence	[ˈsiːkwəns]	n.	顺序,序列
obstacle	[ˈɔbstəkl]	n.	干扰,阻碍
geology	[dʒiˈɔlədʒi]	n.	地质学
glaciology	[ˌglæsiˈɔlədʒi]	n.	冰川学
archaeology	[ˌɑːkiˈɔlədʒi]	n.	考古学

14. Climatic Factors in Plant Growth

Text

All plants have environmental requirements that must be met if they are to thrive. These may be classified broadly as (1) climatic, (2) physiographic, (3) edaphic and (4) biotic. The first is of primary concern here, although the influences of terrain or relief, soil (the edaphic factor), and the interrelations of plants with other plants and with animals cannot be neglected. Climate acts in conjunction with these other factors to set limits to plant growth. Its role is direct in its effects on plants and indirect through its influence on edaphic and biotic factors.

The principal direct effects of climate on plants are exerted by elements of the water and heat budgets, precipitation and soil moisture, humidity, temperature (including soil temperature), sunlight and wind. Variation in one can change the significance of the others in producing different rates of evapotranspiration and photosynthesis. The moisture factors are the most important over large areas of the earth. Water not only goes into the composition of plant cells but also serves as a medium for transport of nutrients to growing cells and through evapotranspiration acts as a temperature control. For most land plants, the immediate source of moisture is the soil. The amount and availability of soil moisture is not necessarily a simple function of precipitation, but is affected by surface drainage conditions and by the ability of the soil to retain moisture as well as by the losses due to evapotranspiration. Thus, swampy areas occur in the midst of deserts and sandy or gravelly soils in rainy climates may be entirely devoid of vegetation. Just as deficient moisture limits plant growth, so excess amounts restrict certain plants by limiting aeration and the oxygen supply on the soil. Excessive soil moisture tends to develop unfavorable soil characteristics and to increase disease damage.

The humidity of the air in which plants grow has varying significance depending upon the type of plant as well as upon the soil moisture available to it. Low vapor pressure of the air induces increased losses of moisture through transpiration. Many plants can withstand low humidities as long as their roots are supplied with adequate moisture. The xerophytic vegetation of dry climates is adapted to limited moisture in several ways. Thick waxy bark and leaves inhibit loss by transpiration. Certain plants have root systems that extend deep into the soil as well as over wide radius to gather moisture from a large volume of soil. Some plants of the desert, notably certain species of cacti, store water during relatively wet periods to be used in dry times. Many desert annuals avoid the prolonged dry spells by passing rapidly through the life cycle from seed to seed after a rain. These are better classified as drought-escaping than drought-adapted.

Whereas moisture provides the medium for the processes of plant growth, heat and light provide the energy. Plants can grow only within certain temperature limits, although the limits are not the same for all plants. For each species and each variety, there is a minimum below which growth is not possible, an optimum at which growth is best, and a maximum beyond which growth stops. Most plants cease growth then the soil temperature drops below about 5℃. If the soil temperature is low, the rate of intake of moisture through the roots is decreased and the plant may not be able to replace water lost by transpiration. Freezing temperatures can thus damage the plant cells by producing chemical changes and desiccation. Alternate freezing and thawing are especially damaging. Species differ a great deal in their adaptation to temperature conditions, however, and many plants can endure long periods of below-freezing temperatures although they do not grow. The effect of high temperature is generally to speed up the growth processes. Under natural conditions, high temperatures are rarely the direct cause of death in plants. Rather, the increased evapotranspiration induced by the heat causes dehydration of the plant cells. Up to a point this can be forestalled if the moisture supply is adequate. Consequently, the temperature-moisture relationships are as important as temperature alone.

The moisture requirements of plants become higher as the energy supply increases, and if moisture is available productivity increases. A given amount of precipitation may result in a moisture deficit in a hot climate, but under lower temperatures the same amount may exceed potential evapotranspiration and create a moisture surplus. If the demands of evaporation from soil and transpiration from plants are not met, wilting and eventual death occur. Both evaporation and transpiration are cooling processes that tend to offset the effects of high temperature. Hot winds increase potential water loss and hasten damage to plant tissues; they may be disastrous even when soil moisture is abundant. As growing conditions become cooler, especially at higher altitudes or latitudes, species are successively eliminated from vegetation formations and ultimately all plant life is prohibited regardless of water in snow, ice, or the frozen soil.

Sunlight supplies energy for photosynthesis and is absorbed by chlorophyll. Without adequate light, green plants fail to develop properly, although some species are adapted to shaded conditions. Light regulates the time required for certain species to flower and produce seed. Ordinarily the amount of light available to growing plants is sufficient for normal development so that it is not of major importance in the geographical distribution of vegetation. However, for an individual plant in a specific environment, light conditions can be critical. Sunshine also is closely related to temperature. In combination with atmospheric heat, absorption of direct radiation can drive the temperature above the maximum limit permitted by the moisture supply. In moderate intensity, radiation can stimulate plants to their optimum development. The albedo of plant surfaces aids the control of leaf and stem temperatures.

Wind influences vegetation directly by its physical action upon plants and indirectly by accelerating moisture loss. Through convective heat transfer it affects plant temperatures. Plant leaf temperatures are modified by increases in wind at low speeds. As speed increases, the proportionate effect is diminished. The rending and tearing action of high-speed or gusty winds is a familiar process. Trees are blown over or stripped of leaves and branches; leaves of bushes are shredded, stems of plants are twisted or broken. An erasion by windblown sand, gravel, or ice particles can also be quite damaging to plants. These are however, local effects. There is no direct worldwide correlation of wind belts with the vegetation pattern. Extensive wind damage is common on exposed mountains, where it combines with the effects of poor soil, low temperatures, and ice or snow cover to create vertical life zones.

Although mean values of the climatic elements may have broad application in determining the suitability of an area for plant growth, the variations from the normal and the extremes are frequently vital considerations. Occasional droughts, floods, heat waves, or frosts can prove fatal to plants otherwise adapted to the normal conditions. A succession of two or three unfavorable years is particularly devastating. Only those plants which escape the disaster or which are able to withstand it survive. Hence climatic anomalies play a leading part in plant adaptation and natural selection.

Duration of minimum conditions for vegetative growth is another consideration which involves the seasonal distribution of the climatic elements. Whether precipitation is well distributed throughout the year, concentrated in a short season, or is erratic sets broad limits for plant associations. Thus, moisture-loving species of the rainy tropics are not adapted to the hot, dry season of the dry summer subtropics. The duration of temperature conditions is likewise restrictive or permissive relative to the plant association in question. Length of growing season is usually defined as the period between the last killing frost of spring and the first killing frost of autumn. It has its widest usage in connection with crop plants, but short growing seasons set limits to natural vegetation as well as pointed out in the discussion of climates dominated by polar air masses. The longer daily duration of sunlight in summer at high latitudes intensifies the growing season so that plants are able to concentrate their annual growth into a shorter period than at lower latitudes. Furthermore, plants differ in their susceptibility to frost damages; some are simply hardier than others.

New Words

thrive	[θraiv]	vi. (植被)繁茂
physiographic	[ˌfiziˈɔɡrəfik]	a. 自然地理学的
edaphic	[iˈdæfik]	a. 土壤的
biotic	[baiˈɔtik]	a. 生命的,生物的
terrain	[ˈterein]	n. 地面,地带
relief	[riˈliːf]	n. 地形,地势
evapotranspiration	[iˌvæpəuˌtrænspiˈreiʃən]	n. 土壤水分蒸发蒸腾损失总量
photosynthesis	[ˌfəutəuˈsinθəsis]	n. 光合作用
drainage	[ˈdreinidʒ]	n. 排水
swampy	[ˈswɔmpi]	a. 沼泽
gravelly	[ˈɡrævəli]	a. 砂砾多的
midst	[ˈmidst]	n. 中间,中央
devoid	[diˈvɔid]	a. 无…的,缺…的
deficient	[diˈfiʃənt]	a. 不足的
aeration	[ˌeiəˈreiʃən]	n. 通风,通气
induce	[inˈdjuːs]	vt. 导致
withstand	[wiðˈstænd]	vt. 抵挡
xerophytic	[ˌziərəˈfaitik]	n. 旱生植物
waxy	[ˈwæksi]	a. 蜡似的
radius	[ˈreidjəs]	n. 范围
bark	[baːk]	n. 茎
cacti	[ˈkæktai]	n. 仙人掌类植物
annual	[ˈænjuəl]	n. 一年生植物
intake	[ˈinteik]	n. 吸入

desiccation	[ˌdesiˈkeiʃən]	n.	干燥
thaw	[θɔː]	vi.	解冻,融化
species	[ˈspiːʃiz]	n.	种类
dehydration	[ˌdiːhaiˈdreiʃən]	n.	胶水
deficit	[ˈdefisit]	n.	欠缺
surplus	[ˈsɜːpləs]	n.	剩余,过剩
wilt	[wilt]	n.	(草木)枯萎,凋谢
offset	[ˈɔːfset]	n.	抵消
tissue	[ˈtisjuː]	n.	组织
disastrous	[diˈzɑːstrəs]	a.	引起灾难的
eliminate	[iˈlimineit]	vt.	淘汰
ultimate	[ˈʌltimit]	a.	最后的,最终的
chlorophyll	[ˈklɔːrəfil]	n.	叶绿素
modify	[ˈmɔdifai]	vt.	调节,缓和
rend	[rend]	vt.	使分列,割裂
shred	[ˈʃred]	vt.	扯碎
abrasion	[əˈbreiʒən]	n.	擦伤
fatal	[ˈfeitl]	a.	命中注定的,必然的
vital	[ˈvaitl]	a.	紧要的
duration	[djuəˈreiʃən]	n.	持续
erratic	[iˈrætik]	a.	不规律的
restrictive	[risˈtriktiv]	a.	限定的
permissive	[pə(ː)ˈmisiv]	a.	随意的
susceptibility	[səˌseptəˈbiliti]	n.	敏感性

15. The Tropical Ocean and Global Atmosphere Project

Text

Introduction

Perhaps the most prominent and coherent signal in the inter-annual variability of the global atmosphere is the Southern Oscillation, an irregular cycle in the pattern of surface pressure, wind and rainfall of the tropical belt, which is particularly marked over the Indian and southern Pacific Oceans. The Southern Oscillation Index(SOI) represents the Tahiti minus Darwin surface pressure values. A positive SOI, when pressure is high over the eastern Pacific and low over the Australian/Indonesian region, maintains the prevailing easterly trade winds. However, in a cycle of period between three and seven years this pattern is reversed: there is a negative SOI with pressure rising over the western Pacific and the easterlies weakening, or even becoming westerlies. The Southern Oscillation is highly correlated with anomalies in sea-surface temperature (SST), a negative SOI being accompanied by a warming of the eastern Pacific Ocean, a deepening of the thermocline and the suppression of upwelling cold, nutrient-rich water off the South American coast, a phenomenon long known as El Nino.

The connection has led scientists recently to refer to such events as manifestations of the El Nino/Southern Oscillation (ENSO). ENSO events are correlated with seasonal temperature anomalies over continental North America during the following season. A downswing in the SOI tends to precede dry monsoon seasons over India by a season or more, and there is evidence that the SOI is also connected with SST anomalies in the Atlantic Ocean. Furthermore, SST anomalies in the extra-tropical Atlantic have been associated with droughts in north-eastern Brazil and the Sather region of Africa, and with cold easterly regimes in winter over western Europe.

Additionally, field experiments, theoretical studies and ocean model simulations have all pointed to an equatorial Kelvin wave as being the link between the Southern Oscillation and El Nino, and have opened up a possibility of forecasting the start of El Nino perhaps three months ahead.

TOGA Objectives

In the light of progress since the Tropical Ocean and Global Atmosphere Project (TOGA) started, the original objectives have been sharpened into two main scientific trusts on different time scales:

Trust I — To develop an operational capability to predict dynamically the evolution of the coupled tropical ocean-atmosphere system beginning with the current state (time-averaged anomalies forecast for up to several months in advance).

Trust II — To explore the predictability of climatic variations of the tropical ocean-global atmosphere system on time scale from one to several years and to understand the processes and mechanisms underlying the variability.

Thrust II is exploratory in nature and will be addressed largely by diagnostic studies so as to determine which physical processes control the longer-term evolution of climatic fluctuations such as the succession of ENSO cycles and the sequence of wet and dry years in the Sahel.

Encouraged by the insight obtained through coupled tropical ocean-atmosphere models (especially over the Pacific), scientists are actively pursuing a detailed plan of action for Trust I. With new satellite measurements of sea-level, wind stress, SST and surface radiation fluxes becoming available, together with the expected rapid progress on developing coupled ocean-atmosphere models, there are grounds for cautious optimism that the objective set by Trust I will be achieved within the lifetime of TOGA.

TOGA Observing Systems

Stated in its simplest terms, the SST pattern determines the distribution of atmospheric convection over the ocean and, therefore, the location of the main heat and moisture sources drive the atmospheric circulation. In turn, the winds force the upper layers of the oceans and thus largely determine the SST distribution, so completing the interactive cycle. An essential task of TOGA is to investigate the nature of these interactions and reproduce them in suitable models. Such models must adequately incorporate the physical processes (especially those operating at the ocean-atmosphere interface), and for this a special program of marine observations will be needed during the TOGA period.

For atmospheric data, TOGA relies on the existing WWW surface and upper-air station networks and on the basic system of two polar-orbiting and five geostationary meteorological satellites. But in addition, TOGA requires enhanced satellite cloud-tracer wind retrievals and upper-wind soundings in the tropics and augmented and/or improved marine meteorological observations from Voluntary Observing Ships and drifting buoys.

For oceanic data, TOGA is leading to a growth of operational oceanography based on the further development of IOC's Global Sea-level Observing System (GLOSS) and IGOSS, since changes in thermal patterns and currents are reflected in sea-level variations that superimpose themselves on the tidal cycle of a much shorter period.

Useful progress has been made in supplementing the WWW by deploying drifting buoys in the southern oceans (over 150 so far) and by installing automatic observing equipment on a number of islands close to the Equator in the Pacific. Oceanic observing systems depend on the use of a combination of old and new techniques based on in situs measuring devices and remote sensing by satellites. Progress in oceanic measurements has been dramatic, particularly in the Pacific.

In the absence of direct measurements of surface fluxes, reliance will have to be placed on quantities derived from atmospheric and surface variables analyzed by operational numerical weather prediction centres. An important recent development has been the agreement by the ECMWF to act as a level III—a (processed real-time data) analysis centre for the WCRP and to achieve the derived fields of surface fluxes, together with all other relevant atmospheric variables.

Coupled Ocean-Atmosphere Models

Remarkable progress is being made in developing coupled ocean-atmosphere models for the tropics and putting them to operational use on a quasi-real-time basis. As an example of the latter, oceanographers at the Climate Analysis Centre of the National Meteorological Center in Washington D.C. use a Cyber 205 supercomputer to run a dynamical model of the tropical Pacific circulation each month. Numerical simulations have already demonstrated considerable skill in hindcasting much of the upper ocean variability observed between 1985 and 1987. Seasonal sea-level variations at a large number of island and coastal stations were accurately reproduced when the model was forced with the actual monthly mean winds. Quasi-real-time simulations are routinely produced using monthly mean wind stresses computed from marine surface data available on the GTS. Comparisons of model results with in situ data already show fair agreement. Weaknesses are most evident in the near-surface layers where problems that remain are associated with the mixed-layer physics, the large uncertainties as to the surface fluxes of heat and momentum and underestimation of higher frequency fluctuations filtered out in the monthly mean forcing fields. Experiments assimilating SST and XBT information to correct some of these problems have met with encouraging results.

A critical and much stronger test than reproducing the variations in sea-level and ocean circulation will be to reproduce the correct surface and near-surface sea temperatures and their variations. For this, the models will have to produce realistic mixed layers and thermoclines as well as changes in thickness associated with changing wind stress and seasonal changes in insolation. A major problem will be to provide atmospheric forcing fields with sufficient accuracy. This underlines the need in TOGA for mutually supporting progress in observations, modeling and process studies. No breakthrough can be expected if any of the three components is seriously deficient.

Conclusion

The TOGA programme can be implemented with only modest new investment, building on existing meteorological and oceanographic systems and mechanisms that already serve the international community. However the incremental resources needed are vital if the programme is to succeed, and that gives all nations the opportunity to make a significant contribution to this unique global endeavour.

New Words

coherent	[kəu'hiərənt]	a. 相干的
positive	['pɔzətiv]	a. 正的
reverse	[ri'və:s]	vt. 使颠倒,使倒转
easterlies	['i:stəlis]	n. 东风带
thermocline	['θə:məˌklain]	n. 温跃层,斜温层
suppression	[sə'preʃən]	n. 抑制,控制
deepen	['di:pən]	vt. 加深,加重
upwelling	[ʌp'welin]	n. 涌升
manifestation	[ˌmænifes'teiʃən]	n. 表现,显现
downswing	['daunswin]	n. 下降趋势
precede	[pri(:)'si:d]	vt. 在…之先,居先于

extra-tropical	[ˈekstrə-ˈtrɔpikl]	a. 温带的
regime	[reiˈʒi:m]	n. 状态
theoretical	[θiəˈretikəl]	a. 理论上的
simulation	[ˌsimjuˈleiʃən]	n. 模拟
wind stress	[waind][stres]	风应力
cautious	[ˈkɔːʃəs]	a. 留意, 小心的
interactive	[ˌintəˈæktiv]	a. 相互作用, 相互影响的
tracer	[ˈtreisə]	n. 追踪物
WWW (World Weather Watch)	[wəːld][ˈweðə][wɔtʃ]	世界天气监测网
geostationary	[ˌdʒi(ː)əuˈsteiʃənəri]	a. 对地静止
buoy	[bɔi]	n. 浮标
operational	[ˌɔpəˈreiʃnl]	a. 运转的
tidal	[ˈtaidl]	a. 潮汐的
superimpose	[ˈsjuːpərimˈpəuz]	vt. 附加
supplement	[ˈsʌplimənt]	n. 增补
deploy	[diˈplɔi]	vt. vi. 附加
install	[inˈstɔːl]	vt. 安装, 设置
situs	[ˈsaitəs]	n. 地点, 部位
remote	[riməut]	a. 遥远的
relevant	[ˈrelivənt]	a. 有关的, 适当的
assimilate	[əˈsimileit]	vt. 同化
implement	[ˈimplimənt]	vt. 执行, 履行
incremental	[inkriˈmentəl]	a. 增加的, 增大的
endeavour	[inˈdevə]	n. 努力, 尽力

16. GOES Data and Nowcasting

Text

GOES (Geostationary Operational Environmental Satellite) is the current operational geostationary meteorological satellite in use by the United States. The heart of the GOES-satellite is its Visible and Infrared Spin Span Radiometer (VISSR) which senses equivalent black body temperature of the scene beneath it with a spatial resolution of 8 km both day and night, as well as 1km resolution visible images during the day, both on a nominal half hourly basis. GOES/VAS is similar to the predecessor GOES with one important exception——its VISSR can also act as an Atmospheric Sounder, thus the acronym is GOES/VAS. The sounder on GOES/VAS has seven CO_2 channels, three water vapor channels and two window channels.

The geostationary satellite has the unique ability to frequently observe the atmosphere (sounders) and its cloud cover (visible and infrared) from the synoptic scale down to the cloud scale. This ability to provide frequent, uniformly calibrated data sets from a single sensor over a broad range of meteorological scales places the geostationary satellite at the very heart of both the understanding and nowcasting of mesoscale weather development. The clouds and cloud patterns observed in a satellite image of animated series of images represent the integrated effect of ongoing dynamic and thermodynamic processes in the atmosphere. When that information is combined with more conventional data such as radar, surface and upper-air observations, then many of the important processes in mesoscale weather development and evolution may be better analyzed and understood. It is from this better analysis and understanding of mesoscale processes that improved nowcasts will become possible.

There are many yet to be answered questions about why the clouds or cloud patterns observed in satellite imagery appear as they do, or are evolving in a particular manner. However, that behavior and evolution is precisely the concern of a major portion of a nowcast. Understanding what the satellite image or image sequence is telling us is the key to developing accurate nowcasts. From the geostationary satellite we have an observation at least every half hour of the earth's cloud cover. These clouds and cloud patterns are the results of specific processes in the atmosphere. It is our task as meteorologists to understand these processes and extend them into the future.

Let's examine what at first might appear to be a fairly simple nowcast problem: the future disposition of early morning fog and stratus. What are the effects of the fog and stratus? In the cloud-free areas, the sun's energy may freely heat the ground and the air, while the cloudy areas are kept several degrees cooler, due mainly to the cloud's higher albedo. This cloud versus no cloud obviously effects surface temperature evolution; less obvious is its effect on afternoon cloudiness, and under proper conditions convective shower development, as well as how rapidly the stable cloud will dissipate in various areas. Dissipation was found to occur from the outside edges of the cloud to region inward, in part due to mixing as a result of differential heating along the cloud boundary. Additionally, brighter cloud areas were found to dissipate later than less bright regions. Using these findings, Gurka (1970) developed a methodology for using information in visible satellite image to predict the time of fog dissipation at any point within the fog area.

There are many factors that may influence surface air temperature. Among them are lapse rate, surface wind speed, surface characteristics and cloud cover. Within the fog and stratus region, the satellite measured brightness should correlate inversely with the amount of sun light reaching the ground, and thus indirectly with temperature. Furthermore, in the cloud-free areas the satellite's infrared data can be used to detect surface temperature gradients. Proper combinations of these pieces of information with direct measurements of surface air temperature should allow for a more precise analysis of surface temperature. This type of initial analysis is a fundamental part of the nowcast. Temperature evolution over the next few hours is a more complicated problem. However, knowledge of when fog and stratus will dissipate over a given area using the previously mentioned work of Gurka (1978), as well as lapse rate, surface characteristics and expected cloud cover would certainly enter into this problem's solution.

Given an area of early morning fog and stratus, where should one expect the first convective clouds and/or showers to form later in the day? It is well known that the static stability of the lower portion of the troposphere is in large part controlled by differential heating. This heating in turn affect the depth of the mixed layer, which partly determines whether or not convective clouds will form. Purdom and Curka (1974) discussed the effects of early morning cloud cover on afternoon thunderstorm development in a weakly forced (synoptically) atmosphere. The situation was likened to that of the land-sea breeze, with the first showers forming in the clear region near the boundary of the early morning cloud cover. Additionally they found that the slower heating rate in early cloudy areas helped to keep those regions free from convection or most of the day.

An accurate and timely mesoscale surface analysis is one of the most important tools available to the forecaster in detecting phenomena that will lead to convective development and intensification. Unfortunately, most detailed surface mesoscale analyses are done for poststorm research, require detailed observations from special mesoscale networks, and are too time-consuming and involved for real time forecasting. When applied in real time situations they often fail because of the lack of a reporting station at the right place at the right time. However, now with GOES data, we have a "reporting station" every 1km using the information in visible data, and every 8 km using the infrared data. Purdom (1976) pointed out that many of the mesoscale phenomena important in the initiation and maintenance of convection, such as the sea breeze, dry lines, lake breezes, area of pre-squall line development, areas of convective cloud merger, and mesoscale high pressure systems, which the forecaster previously tried to infer from mesoscale patterns are readily detectable in GOES image.

Not all convective triggering mechanisms have received as much attention as those which result in severe thunderstorms. This is especially true of thunderstorm development in a weakly forced atmosphere. In a recently completed study using GOES satellite data, generation mechanisms were classified into four categories: (1) merger-development on an arc cloud as it moved into a cumulus region; (2) intersection-development where two clouds come into contact; (3) local-forcing-development due to some local mechanism (not 1 or 2 above) such as a sea breeze, front, river breeze or other local generating mechanisms; (4) undeterminable those whose generation mechanism could not clearly be determined such as new storms from beneath a cirrus deck. These results show quantitatively that while early in the day local forcing dominates the convective generation mechanism, later in the day when the most intense has developed, the dominant generation mechanisms are mergers and intersections.

On selected days during the past few convective seasons, the National Earth Satellite Service (NESS) has operated the GOES system in a special three-minute interval imaging mode. Radar PPI images over the convective areas were also collected on these days. This unique data set is allowing meteorologists for the first time to observe the development of deep convective storms with a spatial and temporal resolution compatible with the scale of the mechanisms responsible for their triggering. Movies made from these data show that convective scale interaction is of primary importance in determining the evolution of deep convection. In fact, thunderstorm evolution that may appear as random in nature with radar is often observed as very well ordered when viewed with GOES image.

The question of how to best use satellite and radar information together for nowcast purposes is still far from resolved. Most commonly, these two data sets have been analyzed separately, then mentally merged by the meteorologists. Little work has been done in extracting quantitative information about mesoscale convective features and individual thunderstorms by combining digital information from satellite and radar. Papers by Adler and Fenn (1979) and Adler (1981), in which thunderstorm intensity determined from 5 minute interval GOES data are examined in relation to simultaneous radar measurements are certainly initial steps in this direction. However, these efforts are more of a comparison of the two different data types (certainly necessary) rather than combination. This area has tremendous potential for further development. Both satellite and radar provide independent measures of thunderstorm intensity. Satellite data provides such information as cloud top mean vertical growth and anvil expansion rates, while radar data provides information about reflectivity, volumetric echo properties and their changes in time. Perhaps combining these pieces of information in a simple diagnostic cloud model will lead to an intensity measure for thunderstorms.

Geostationary satellite data must become one of the major footings upon which mesoscale forecasting problems of the future are based. By combining satellite data with more conventional data such as radar and surface observations, many of the features important in mesoscale weather development and evolution may be better analyzed and understood. This better analysis and understanding of mesoscale processes is necessary if short range forecasting is to be successful.

New Words

geostationary	[ˌdʒi(:)əuˈsteiʃənəri]		a. 对地静止
VISSR visible and Infrared Spin Scan Radiometer	[ˈvizəbl][ˈinfreˈred] [spin][skæn][ˌreidiˈɔmitə]		可见光和红外自旋扫描辐射仪

equivalent	[iˈkwivələnt]	a. 等值,等价
spatial resolution	[ˈspeiʃəl][ˌrezəˈljuːʃən]	空间分辨率
nominal	[ˈnɔminl]	a. 名义上的
acronym	[ˈækrənim]	n. 首字母缩略词
sounder	[ˈsaundə]	n. 探测器
calibrate	[ˈkælibreit]	vt. 校准,使标准化
sensor	[ˈsensə]	n. 传感器
animated	[ˈænimeitid]	a. 栩栩如生的
thermodynamic	[ˌθəːməudaiˈnæmik]	a. 热力学的,热力的
meso-scale	[ˈmesəskeil]	中尺度
evolution	[ˌiːvəˈluːʃən]	n. 演变,发展
evolve	[iˈvɔlv]	vt. 发展
sequence	[ˈsiːkwəns]	n. 顺序
stratus	[ˈstreitəs]	n. 层云
versus	[ˈvəːsəs]	prep. 与…相对
lapse rate	[læps][reit]	递减率
timely	[ˈtaimli]	a. 合时的,正好的
consume	[kənˈsjuːm]	vt. 浪费,消耗
squall line	[skwɔːl][lain]	飑线
merger	[ˈməːdʒə]	n. 合并,结合
trigger	[ˈtrigə]	vt. 触发
arc	[aːk]	n. 弧
contact	[ˈkɔntækt]	n. 接触
compatible	[kəmˈpætəbl]	a. 协调的,不矛盾的,相容的
primary	[ˈpraiməri]	a. 主要的
quantitative	[ˈkwɔntitətiv]	a. 定量的
simultaneous	[ˌsiməlˈteinjəs]	a. 同时发生的,同时做的
anvil	[ˈænvil]	n. 云砧
conventional	[kənˈvenʃənl]	a. 平常的,常规的

17. Acidic Deposition

Text

The term "acid rain" was first used by Robert Angus Smith in 1872 to describe the acidic nature of the rain falling around Manchester in one of his early reports as the first Chief Alkali Inspection of the UK. Acid rain is the popular term for a very complex environmental problem. Over the last 15 years, evidence has accumulated on change in aquatic life and soil pH in Scandinavia, Canada and the northeastern United States. Many believe that these changes are caused by acidic deposition traceable to pollutant acid precursors that result from the burning of fossil fuels. Acid rain is only one component of acidic deposition, a more appropriate description of this phenomenon. Acidic deposition is the combined total of wet and dry deposition, with wet acidic deposition being commonly referred to as acid rain.

The mechanism of wet deposition differs considerably from that of dry deposition because the rates of dry deposition are directly dependent on pollutant concentration velocity and deposition velocity which, in turn, depend on the nature of the uptake or receiving surfaces on the land or the sea. By contrast, rates of net deposition do not depend on the underlying surface characteristics but on the precipitation rate, the washout ratio and the ambient air concentration.

Acidity is defined in terms of the pH scale, where pH is the negative logarithm of the hydrogen ion $[H^+]$ concentration: $pH=-\log[H^+]$. The best starting point to consider the acid-forming reactions of cloud-water is to ask the question: "What would be the pH of cloud-water if there was no atmospheric pollution?" If one leaves aside the possible contributions of sulphur dioxide and oxides of nitrogen from volcanoes, swamps and lightning strikes, etc., then the major "natural" gas in the atmosphere to contribute acidity would undoubtedly be carbon dioxide. In the simplest case, CO_2 dissolves in raindrops forming carbonic acid. At a temperature of 20℃, the raindrops will have a pH of 5.6. The value often labeled as that of clean or natural rainwater. It represents the baseline for comparing the pH of rainwater which may be altered by SO_2, or NO_x oxidation products. Figure 1 illustrates the pH scale with the pH of common items and the pH range observed in rainwater. The pH of rainwater can vary from 5.6 due to the presence of H_2SO_4, and HNO_3, dissolved or formed in the droplets. These strong acids dissociate and release hydrogen ions, resulting in more acidic droplets. Basic compounds can also influence the pH. Calcium, magnesium, and ammonium ions help neutralize the rain droplet and shift the overall H^+ toward the basic end of the scale. The overall pH of any given droplet is a combination of the effects of carbonic acid, sulfuric and nitric acids, and any neutralizers such as ammonia.

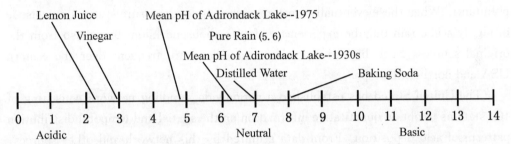

Fig. 1 The pH scale is a measure of hydrogen ion concentration. The pH of common substances is shown with various values along the scale. The Adirondack lakes are located in the state of New York and are considered to be receptors of acidic deposition. Source: United States Environmental Protection Agency. "Acid Rain—Research Summary" 1979

Dry deposition occurs when it is not raining. Gaseous SO_2, NO_2 and HNO_3, and acid aerosols are deposited when they contact and stick to the surface of water bodies, vegetation, soil and other materials. If the surfaces are moist or liquid, the gases can go directly into solution; the acids formed are identical to those that fall in the form of acid rain, forming acids in the liquid surfaces if oxidizers are present. During cloud formation when rain droplets are created, fine particles or acid droplets can act as seed nuclei for water to condense. This is one process by which sulphuric acid is incorporated into the droplets. While the droplets are in the cloud, additional gaseous SO_2, and NO_2, will impinge on them and be absorbed. These absorbed gases can be oxidized, by dissolved NO_2, or other oxidizers, lowering the pH of the raindrop. As the raindrop falls beneath the cloud, additional acidic gases and aerosol particles may be incorporated into it, also affecting its pH.

Natural and man-made emissions are released into the atmosphere by a complicated mixture of physical processes. Apart from volcanoes, most natural and some man-made emissions (e. g. by traffic) originate at ground level whilst others are ejected from stacks often over a hundred metres high. All emissions reach an altitude dependent on the mixing characteristics of the atmosphere prevailing at the time of release.

Pollutants at any height may be carried away (or advected) by winds and therefore climatic factors determine the direction and speed of transport. At the same time dispersion may take place by turbulent eddying. However, there are also weather conditions that cause the accumulation of pollutants. Slow-moving high-pressure areas with low advection and dispersion rates, for example, have the effect of concentrating pollutants. When these eventually meet an advancing low-pressure system, an episode of highly acidic rain may be experienced often a thousand kilometers or so from the original sources. Such high-low pressure transitions are frequent over the eastern USA and northern Europe.

The United Stated has established a nationwide sampling network consisting of 100 stations to obtain quantitative information on the spatial and temporal distribution patterns of acid deposition. From data acquired by this network and other sources, the lowest rainwater pH isopleths are associated with the regions of highest SO_2, emissions. While there is considerable controversy over the quality and strength of the link between SO_2 and NO_x emissions from stationary sources and subsequent acid deposition hundreds of kilometers downwind, the National Research Council has concluded that 50% reduction in the emissions of sulfur and nitrogen gases will produce about 50% reduction in the acids deposited on the land and water downwind of the emission source.

In the eastern United States, acid rain consists of 5% sulfuric acid, 30% nitric acid, and 5% other acids. In the West, windblown alkaline dusts buffer the acidity in rains occurring over many rural areas. Whereas in urban areas 80% of the acidity is due to nitric acid, average pH in rainfall over the eastern United States for the period April 1979—March 1980 was less than 5.0, with some areas less than 4.2.

Aquatic systems in areas of large snowfall accumulation are subjected to a pH surge during the spring thaw. Acidic deposition is immobilized in the snowpack, and when warm springtime temperatures cause melting, the melted snow flows into streams and lakes, potentially overloading the buffering capacity of the aquatic system. Maximum fish kills occur in the early spring due to the "acid shock" of the first melt-water, which releases the pollution accumulated in the winter snowpack. This first melt may be 5—10 times more acidic than rainfall. In Sweden, thousands of lakes are no longer able to support fish. In the United States the number of polluted lakes is much smaller, but many more may be pushed into that condition by continued acidic deposition. In Canada danger to aquatic systems and forest ecosystems is a matter of considerable concern.

A second area of concern is reduced tree growth in forests. As acidic deposition moves through forest soil, the leaching process removes nutrients. If the soil base is thin or contains barely adequate amounts of nutrients to support a particular mix of species, the continued loss of a portion of the soil minerals may cause a reduction in future tree growth rates or a change in the types of trees able to survive in a given location.

The two components of acidic deposition described are wet deposition and dry deposition. The collection and subsequent analysis of wet deposition are intuitively straightforward. A sample collector opens to collect rain water at the beginning of a rain storm and closes when the rain stops. The water is then analyzed for pH anions (negative ions), and cations (positive ions). The situation for dry deposition is much more difficult. The collection of particles settling from the air is very dependent upon the surface material and configuration. The surfaces of trees, plants, and grasses are considerably different from that of the round, open-top canister often used to collect dry deposited particles. After collection, the material must be suspended or dissolved in pure water for subsequent analysis.

The primary constituents to be measured are pH of precipitation, sulfates, nitrates, ammonia, chloride ions, metal ions, phosphates and specific conductivity. The pH measurements will help to establish reliable long-term trends in patterns of acidic precipitation. The sulfate and nitrate information will be related to anthropogenic sources where possible. The measurements of chloride ions, metal ions, and phosphates are related to sea spray and windblown dust sources. Specific conductivity is related to the level of dissolved salts in precipitation.

New Words

aquatic	[ə'kwætik]	a.	水生的,水栖的
traceable	['treisəbl]	a.	起源于…的
precursor	[pri(:)'kə:sə]	n.	前驱,前辈
mechanism	['mekənizəm]	n.	结构
velocity	[vi'lɔsiti]	n.	速度
uptake	['ʌpteik]	n.	吸收

ratio	[ˈreiʃiəu]	n.	比率
ambient air	[ˈæmbiənt]		环境大气
logarithm	[ˈlɔɡəriθm]	n.	对数
ion	[ˈaiən]	n.	离子
neutral	[ˈnju:trəl]	a.	中性的
basic	[ˈbeisik]	a.	碱性的
distilled water	[disˈtild][ˈwɔ:tə]		蒸馏水
vinegar	[ˈviniɡə]	n.	醋
droplet	[ˈdrɔplit]	n.	微滴
calcium	[ˈkælsiəm]	n.	钙
magnesium	[mæɡˈni:ziəm]	n.	镁
ammonium	[əˈməunjəm]	n.	铵
aerosol	[ˈɛərɔsɔl]	n.	气溶胶
particle	[ˈpɑ:tikl]	n.	微粒
eject	[iˈdʒekt]	vt.	发射
dispersion	[disˈpə:ʃən]	n.	传播
isopleth	[ˈaisəupleθ]	n.	等值线
controversy	[ˈkɔntrəvə:si]	n.	争论,辩论
reduction	[riˈdʌkʃən]	n.	减少
alkaline	[ˈælkəlain]	a.	碱性的
buffer	[ˈbʌfə]	vt.	缓和
rural	[ˈruər(ə)l]	a.	乡下的,农村的
surge	[sə:dʒ]	n.	起伏,高涨
thaw	[θɔ:]	n.	(冰、雪等)解冻、溶化
immobilize	[i:məubilaiz]	vt.	使不动,使固定
shock	[ʃɔk]	n.	冲击
ecosystem	[i:kəˈsistəm]	n.	生态系统
leach	[li:tʃ]	n.	沥滤
intuitively	[inˈtju(:)itivli]	ad.	直觉地,直观地
anion	[ˈænaiən]	n.	阴离子
cation	[ˈkætaiən]	n.	阳离子
configuration	[kənˌfiɡjuˈreiʃən]	n.	结构
canister	[ˈkænistə]	n.	罐,茶筒
suspend	[səsˈpend]	vi.	悬浮
conductivity	[ˌkɔndʌkˈtiviti]	n.	传导性
phosphate	[ˈfɔsfeit]	n.	磷酸盐
sulfate	[ˈsʌlfeit]	n.	硫酸盐
chloride	[ˈklɔ:raid]	n.	氯化物
spray	[sprei]	n.	浪花,水花
windblown	[waindˈbləun]	a.	(树等)终年挡风的

18. Weather Modification

Text

Weather modification is any change in weather that is induced by human activity. The activity may be intentional or inadvertent. It has long been known that liquid water drops frequently exist in clouds at temperatures far below 0℃. But it was discovered that these supercooled water droplets could be converted to ice crystals by seeding with dry ice, solid carbon dioxide (CO_2) at a temperature about $-80℃$. Since the coexistence of ice crystals and water drops is one important process leading to growth of cloud particles to precipitation size, this discovery led to revived interest in the possibility of artificial stimulation of precipitation.

Since World War II considerable research has gone into methods of enhancing precipitation by cloud seeding. Cloud seeding is an attempt to stimulate natural precipitation process by injection nucleating agents into clouds. Cloud seeding cannot produce condensation of water drops; clouds must already exist before seeding can have any effect. Most cloud seeding experiments are directed at cold-clouds. The objective of seeding cold clouds is to stimulate the Bergeron Process in clouds that are deficient in ice crystals. The seeding (nucleating) agent is either dry ice or silver iodide (AgI), a substance with crystal properties similar to those of ice. Silver iodide crystals are freezing nuclei that are active at $-4℃$ and below. Within a cloud, dry ice pellets are cold enough to cause surrounding supercooled water droplets to freeze and the frozen droplets then function as nuclei that grow into snowflakes.

In warm clouds of relatively uniform droplet size, sea-salt crystals and other hygroscopic substances can be injected to trigger development of relatively large cloud droplets. Such seeding stimulates the collision-coalescence process.

Clouds targeted for seeding are those rich in supercooled water droplets and deficient in ice crystals. In an attempt to identify the type(s) of cloud most suitable for precipitation enhancement through seeding, scientists employ an aircraft specially outfitted with sophisticated instruments that measure the size and concentration of cloud and precipitation particles. Another air-craft seeds clouds with silver iodide crystals or dry ice pellets, and, on the ground, an array of precipitation-gauge radar and other weather instruments monitor the effectiveness of seeding.

Does cloud seeding work? Over a period of several years in the late 1970s and early 1980s, NOAA scientists conducted a statistically rigorous experiment designed to test the effectiveness of weather modification. The experiment was carried out over southern Florida and involved the seeding of cumulus clouds. The results were very encouraging, showing a 25 percent increase in rainfall on days when silver iodide was the seeding agent, as compared with days when sand was the bogus seeding agent. This finding was statistically significant at the 90 percent level, meaning that there's only a 10 percent probability that the rainfall increase was simply the result of chance.

In some experiments, precipitation actually appears to have been reduced by seeding? A possible explanation for the reduced precipitation is that the clouds were over seeded. Prior to seeding, cumulus clouds apparently contained just enough ice crystals for precipitation. The addition of more nuclei by seeding probably produced too many ice crystals competing for too few supercooled water droplets. Consequently the seeding generated a large number of ice pellets that were so small that they tended to remain suspended in the clouds, or they vaporized before reaching the ground.

The physical evidence gained so far proves that cloud seeding enhanced precipitation when the right kind and amount of material is applied at the right time and place in a cloud. The task of determining what is "right" is very difficult. New and emerging technologies may now enable us to meet the challenges of measuring if and when a complex cloud system is "right", if and how, seeding material can be effectively delivered and whether or not precipitation over an area can or cannot be beneficially increased.

Fog can pose a serious hazard to both surface and air travel. Many auto accidents and ship collisions and aircraft crashes have been attributed, at least in part, to visibility restrictions caused by dense fog. Fog frequently forces flight delays, reroutings and cancellations that cost airlines millions of dollars each year and inconvenience for thousands of passengers. Although the need for an effective method of fog dispersal is great, especially at airports, little progress has been made in this area for both technical and economic reasons. During World War II, the British had considerable success in clearing warm radiation fogs from runways at 15 military airfields. Heat from fuel burners deployed alongside runways raised the air temperature, thereby lowering the relative humidity to below saturation so that fog droplets vaporized.

With shallow radiation fogs at airports, some clearing may be achieved by using helicopters to induce vertical circulation of air. The rotor blades force the warmer unsaturated air that is just above the fog layer to mix with it. Humidity of the mixture falls below 100 percent, and fog droplets vaporize. Another approach to fog dispersal applies techniques of cloud seeding. Warm fogs are seeded with hygroscopic substances that absorb water vapor, thereby reducing the relative humidity by reducing the vapor pressure. Fog droplets vaporize, and hygroscopic droplets grow into raindrops that fall to the surface. Cold fogs (10 to $-20°C$) are seeded with dry ice pellets that stimulate the Bergeron process.

Efforts to suppress hail have deep historical roots. Indeed, in fourteenth-century Europe, church bells were rung and cannons fired in the belief that the attendant noise would somehow ward off hail. Today scientists fire silver iodide crystals into thunderclouds. They theorize that silver iodide crystals stimulate the formation of large number of small hailstones, which will melt long before they reach the ground, instead of normal development of small numbers of larger hailstones, which could devastate crops. Methods other than seeding have been tried to induce freezing of the supercooled water droplets in hail-producing cumulonimbus clouds. Shock waves produced by explosive charges of TNT carried into the clouds by rockets have been reported to have significantly reduced hail damage. Perhaps the annoying sonic boom of supersonic aircraft might become useful for this purpose.

When supercooled water droplets are induced to freeze, latent heat of fusion is released, which can help to intensify the convection within the cloud. If the convection pattern within a storm could be modified, then it is conceivable that horizontal circulation might also be changed. Although the spiral bands of clouds and precipitation extend hundreds of kilometers from the eye of a hurricane, the most intense upward air motion occurs in a relatively small ring surrounding the eye. This relatively narrow zone of great towering clouds appears to act like "chimney" of the storm, the heat of condensation within it providing much of the "draft" that sucks the air inward at lower levels from hundreds of kilometers away. If this chimney could be widened, then the intensity of the draft and therefore the wind velocities might be diminished. Silver iodide, injected into the wall of clouds adjacent to the eye of the storm, may convert supercooled water droplets to ice crystals releasing heat of fusion. The net effect might be to reduce the pressure gradient: the size of the storm would be increased, but its intensity decreased. A few cloud seeding experiments on hurricanes have already been carried out (by Project Stormfury under NOAA and the U. S. Navy) with this idea in mind. Although some changes in clouds and the wind pattern were detected in a couple of instances, it is as yet too soon to draw any conclusions about the efficiency of the technique.

Although we are still not certain about how lightning is generated, there is a great deal of interest in learning how to suppress it. Lightning strokes damage millions of dollars worth of forests and property each year. Experimental results indicate that clouds that have been seeded by silver iodide have about one-third fewer cloud-to-ground lightning strokes than those that are unseeded. No one is sure why increasing the number of ice crystals should have this effect.

Research in the National Oceanic and Atmospheric Administration (NOAA) Federal-State Cooperative Program in Weather Modification Research is focused on four key subsets of problems: (1) determination of the presence, persistence, and natural utilization of supercooled liquid water, (2) determination of potentials and methods for effective delivery of the seeding material to the supercooled liquid water, (3) verification of the effects of seeding, and (4) quantification of benefits of any increased precipitation or suppressed hail. Satisfying proof of our hypotheses in these areas will come only with measurements that provide direct physical evidence.

New Words

modification	[ˌmɔdifiˈkeiʃən]	n. 控制,影响
intentional	[inˈtenʃənəl]	a. 有意的,故意的
inadvertent	[ˌinədˈvəːtənt]	a. 无意中的
revive	[riˈvaiv]	vt. 恢复,使复兴
nucleate	[ˈnjuklieit]	vt. 使成核
agent	[ˈeidʒənt]	n. 药剂,催化剂
deficient	[diˈfiʃənt]	a. 不足的,缺乏的
pellet	[ˈpelit]	n. 小球
snowflake	[ˈsnəufleik]	n. 雪片
hygroscopic	[ˌhaigrəuˈskɔpik]	a. 吸湿的,收湿的
seeding	[ˈsiːdiŋ]	n. 播散,催化
rigorous	[ˈrigərəs]	a. 严格的,精密的
dispersal	[disˈpəːsəl]	n. 散开,驱散
deploy vi	[diˈplɔi]	vt. 散开,疏散开
saturation	[ˌsætʃəˈreiʃən]	n. 饱和

rotor	['rəutə]	n. 电机的转子
attendant	[ə'tendənt]	a. 伴随的
ward	[wɔːd]	vt. 防止
sonic	['sɔnik]	a. 声音的
boom	[buːm]	n. 隆隆声,轰轰声
conceivable	[kən'siːvəbl]	a. 可能的,可以想到的
verification	[ˌverifi'keiʃən]	n. 证实

19. Agrometeorological Forecasting

Text

Scope of Agrometeorological Forecasting

Agrometeorological forecasting is concerned with the assessment of current and expected crop performance, including dates of crop-development stages (especially maturity) and yields (quantity and quality), and other factors affecting production patterns, such as densities of sowing and locations and acreages of planting. This field of forecasting must be distinguished from the weather forecasting for agriculture, which deals with special forecasts of weather elements affecting farm operations, e. g. , forecasts for spraying and for estimating probabilities of occurrence of potential hazards (frost, fire, hail, severe rainfall).

Agrometeorological forecasting may be called crop prediction without weather forecasting, since it uses actual past and present meteorological data (and not their values extrapolated into the future) to predict crop performance in the future.

For production purposes the most significant agrometeorological forecasts pertain to crop development and ripening phases, crop yields, soil-moisture availability and water supply, wintering conditions and freezing areas of winter crops, rates of irrigation and heat supply.

Basic Principles of Agrometeorological Forecasting

Many of the techniques of agrometeorological forecasting are based on the statistical relationships existing between the dependent variables to be estimated (yields, date of flowering, etc.) and independent agrometeorological variables (index of soil moisture, atmospheric moisture stress). The independent variables or predictors can often be readily chosen by agrometeorological experience or intuition; where several variables are interrelated (e. g. , moisture is affected by temperature) the dominant variables may be chosen by multivariate regression analysis. This is necessarily a practical approach leading to successful predictions, but requiring a largely empirical input. This also implies a fundamental lack of understanding of the total impact of all the environmental factors bearing in a complex way on the crop factors to be estimated. No explanation of cause or effect is therefore provided; such explanation remains a subject for more research. The method requires only a knowledge of the major persistent factors affecting crop performance rather than long-range forecast of their behaviour, which are at the present time not available in sufficient detail and accuracy.

The method is not transferable from one region to another, since agrometeorological variables are spatially homogeneous only over limited areas with the same soil type, topography, climate, cultivation practices, etc. Thus the forecasting relationships (linear or quadratic equations) vary from one region to another.

The underlying principles of such agrometeorological forecasts can be postulated as follows:

(a) Current crop conditions can be assessed from past weather data and determine to some extent the potential yielding ability.

(b) Soil moisture can be estimated from past weather data and is one of the most important environmental variables determining quality and quantity of crop yields.

(c) Current weather conditions have a tendency to persist into the future for a number of days and thereafter tend in a statistical sense to the normal, with a known probability distribution.

(d) The probability distribution of the most important weather elements can be used to determine probable future changes in the current crop condition and its potential yielding ability.

(e) Weather elements, particularly temperature and radiation, are conservative elements as regards to their macro-scale distribution so that weather records, even from a limited number of observing stations, can be used for estimating yields and production over a relatively large area.

Accuracy of Forecasts

As mentioned before, present (i.e., initial) weather conditions determine later plant development stages, growth and yields. Therefore actual observations of weather elements significant for agriculture are very useful for agrometeorological forecasting and provide a basis for more accurate predictions than those based on regular weather forecasts of these same elements, for although valid sometime in the future, long-or, medium-range, weather-forecasting methods have not yet reached the level of accuracy desirable for operational use.

Any deficiencies in the accuracy of agrometeorological forecasting depend on:

(a) how well the initial observations represent regional conditions.

(b) how homogeneous the regional conditions (climate, soil characteristics, etc.) are.

(c) how accurate the observations themselves are, and (d) how sensitive the model is to the variations in the agrometeorological variable being forecast.

Phenological Forecasts

Almost every agrometeorological forecast contains some elements of a phenological forecast, which is a prediction of the dates of occurrence of the main crop-development phases, each of which has different climatic requirements. For example, forecasts of the onset of flowering of fruit trees or the dates of ripening of fruits are important for making management decisions in regard to frost-protection operations, manpower requirements and marketing. In grain protection, a forecarts of the expected onset of the main crop-development phases (e. g. , jointing, heading, flowering, soft dough and ripening) are required in connexion with the assessment of soil-moisture conditions and yields. In principle, two approaches are available for the estimation of these dates from meteorological data: First, by the use of predicted temperatures, based on climatological data, in degree-day type expressions (and their various derivatives) which are related to the rate of plant development; Second, by relating meteorological data or phenological observations obtained earlier in the growing season to the dates of plant/crop-development stages occurring later in the growing season.

Over-Wintering Forecasts

Long-range forecasts of expected over-wintering conditions and of the state of the winter crop in spring are very important in many countries. The forecasts should refer not only to regions expected to have unfavourable winter conditions but also these expected to have poor winter crops requiring partial or complete re-sowing in spring. Damage and destruction to crops during wintering occur for a variety of reasons, chiefly freezing, soaking, accumulating a rather thick crust of ice in the soil, etc.

For regional crop yield and production estimates, it is important to know the expected proportion of damaged or lost crops resulting from unfavourable wintering conditions. Methods of forecasting over-wintering conditions and prognostic equations for estimating the extent of frost-damaged areas in different regions have been developed.

Soil-Moisture Assessments and Forecasts

The forecast or assessment of available moisture in a 1 m layer of soil at the beginning of spring is of great assistance to farm operators and agricultural planning agencies. Such a forecast is usually based on climatological water-balance methods or empirical regression-type equations. An assessment of moisture conditions is based on past and present climatological data (e. g. , precipitation, temperature, wind) with or without the use of soil-moisture measurements. An extrapolation of this current estimate into the near future is possible through the use of long-term averages or other statistical values of the above meteorological data in the water-balance equation. On the other hand, a soil-moisture forecast equation is based on a statistical analysis of recorded soil-moisture data related to one or several other agrometeorological variables. This approach uses, sometimes on a probability basis, the occurrence of events in the past for extrapolation in the near future.

Modeling of Yield Forecasts

Of the agrometeorological forecasts in use, probably the most important economically are agrometeorological forecasts of crop yields. The evolution of these methods has made such rapid strides over the past 10 to 15 years that agrometeorologists have now derived them for the main cultivated crops in a number of countries. The leading countries concerned with yield forecasts include Canada, German, India, Japan, the U. S. and the U. S. S. R. Most operational yield forecasts were developed for annual grain crops because of their major role in world food supply and their economic significance in international trading. However, successful forecast models for soya bean, flax, sugar beet and other commercial crops of regional significance have been reported.

There are at least three approaches to modeling the impact of weather and climate on crop yields:

(a) Crop-growth simulation models describing the detailed impact of meteorological variability on biological/physical processes that occur within a typical plant or plant canopy;

(b) Crop-weather analysis models which are a research tool for the analysis of crop responses to selected agrometeorological variables;

(c) Empirical-statistical models using a sample of yield data from an area and a sample of weather and soil data from the same area to produce estimates of coefficients in the model by some sort of regression technique. The empirical-statistical approach is mostly used in the currently operational crop-yield and production forecasts on a national or regional basis.

New Words

agrometeorological	[ˈægrəuˌmiːtiərəˈlɔdʒikəl]	a. 农业气象的
assessment	[əˈsesmənt]	n. 评价，估定
acreage	[ˈeikəridʒ]	n. 英亩数
spray	[sprei]	n. 喷药
ripen	[ˈraipən]	vi. 成熟
predict	[priˈdikt]	vt. vi. 预言，预报
intuition	[ˌintju(ː)ˈiʃən]	n. 直觉
regression	[riˈgreʃən]	n. 回归
imply	[imˈplai]	vt. 包含，暗示
sufficient	[səˈfiʃənt]	a. 充分的
transferable	[trænsˈfɜːrəb(ə)l]	a. 能转移的
spatially	[ˈspeiʃəli]	ad. 空间地
homogeneous	[ˌhɔməuˈdʒiːnjəs]	a. 均一的，齐次的
topography	[təˈpɔgrəfi]	n. 地形学
cultivation	[ˌkʌltiˈveiʃən]	n. 耕种，耕作
linear	[ˈliniə]	a. 线的，直线的
quadratic	[kwəˈdrætik]	a. 二次的
quadratic equation	[kwəˈdrætik][iˈkweiʃən]	二次方程式
postulate	[ˈpɔstjuleit]	vt. 假定
assess	[əˈses]	vt. 确定，评估
tendency	[ˈtendənsi]	n. 倾向，趋势
conservative	[kənˈsəːvətiv]	a. 保守的
valid	[ˈvælid]	a. 确凿的，正确的
phenological	[ˌfiːnəˈlɔdʒikəl]	a. 物候学的
connexion	[kəˈnekʃən]	n. 联系
derivative	[diˈrivətiv]	n. 导数
partial	[ˈpaːʃəl]	a. 不完全的
prognostic	[prɔgˈnɔstik]	a. 预报的
extrapolation	[ˌekstrəpəuˈleiʃən]	n. 外推法
stride	[straid]	vi. 迈进
soya	[ˈsɔiə]	n. 大豆
flax	[flæks]	n. 亚麻
canopy	[ˈkænəpi]	n. 树冠

20. Radar Measurement of Rainfall Intensity

Text

The quantitative measurement of rainfall is of great importance. The traditional raingauge provides a fairly accurate method of obtaining measurements at a point but there is a greater requirement by meteorologists, hydrologists, communications engineers and others for detailed measurements of the mesoscale pattern of rainfall over extended areas. Often these measurements are needed in real time. It is likely that radar techniques will become the standard method of meeting many of these requirements, partly because of the unique capability of radar to observe areal distribution of precipitation and partly because data from a large area are immediately available at a single centre with a minimum of telemetry requirements. When the subject was reviewed by Kessler (1968), there were still some practical problems in the way of the operational use of quantitative radar techniques. It is only during the last decade that progress in data processing and communications, together with experimental assessments of accuracy, have confirmed the practical utility of radar in this field.

Techniques of measuring rainfall by radar fall into two broad categories. One depends on a relationship between the rainfall rate R and the associated attenuation of the radar energy; the other depends on a relationship between R and the intensity of the backscattered radiation, or radar reflectivity. More sophisticated method, which we shall mention only in passing, combines measurements of attenuation and reflectivity made simultaneously at two wavelengths. This approach has been suggested as a way of measuring not only R but also liquid water content and the median volume diameter of the raindrops.

The simple attenuation method of measuring rainfall exploits the fact that the microwave attenuation by rainfall at wavelengths near 1 cm is related nearly linearly to the rainfall rate. This linear relationship formed the basis of several experiments to measure path-integrated rainfall at a wavelength of 8.6 mm using two or more known targets along a path. The difficulties with this technique are the need for large numbers of remote targets (or receivers) to achieve areal measurements and occasionally the total loss of signal owing to excessive attenuation in heavy rain. Thus, although this method is potentially useful for path-integrated measurements in nontorrential rain, the approach which has been generally adopted as the basic radar-rainfall technique exploits instead the relationship between rainfall rate and radar reflectivity. Because of its practical importance, we shall now consider this latter approach in some detail.

The radar reflectivity approach of measuring rainfall involves the measurement of the radar energy scattered back to the radar aerial by the raindrops within a number of pulse volumes, each one defined as in Figure 1 (omitted) by the radar beamwidth, the range and the pulse length. Measurements are made simultaneously in pulse volumes at many ranges out to 100 km or more and typically at different azimuths as the beam rotates about a vertical axis. It has been shown that, provided the pulse volume is uniformly filled by rain, then

$$P_r = \frac{C_r c_p K Z}{r^2} \tag{1}$$

where P_r is the average power returned from the rain at range r, C_r is a function of the known radar parameters, c_p is approximately a constant related to the dielectric properties of the raindrops. K is the coefficient of attenuation (due mainly to rain) as the radiation traverses to and from the range of interest, and Z is the radar reflectivity factor, defined as the summation per unit volume of the sixth power of the drop diameters. The simple sixth-power relationship is a direct consequence of the Rayleigh scattering law which applies as long as the radar wavelength is long compared with the diameter of the scattering particles. This is not valid for radar scattering from large hail, but it does hold good for rain at the wavelengths normally used in weather radars.

The reflectivity factor Z in Equation (1) is related to the rain fall rate through the equation, $Z = aR^b$, where R is the rainfall rate and a and b are empirical constants. While the measurement of rainfall outlined above is straightforward in concept, there are a number of difficulties which call for careful experimental design. In addition to some natural variability in the "constants" a and b, there are problems arising from the fluctuating character of the radar echo from rain and from the fact that the pulse volume at long range is of a considerable distance above the ground. These and related problems are considered next.

The radar signal from rain is the sum of the signals from all the raindrops distributed and moving at random in each pulse volume. The result is a strongly fluctuating signal for which the average power contains information about the rainfall intensity while the fluctuations contain information about the radial motion of the raindrops. The fluctuations provide the basis of a ground-clutter discrimination method. However, the very existence of such fluctuations means that the average must be obtained with some care. Integration can be carried out over samples of P_r that are separated in range or time (or both). The need for appropriate integration is well understood, however, and errors arising from inadequate integration are usually small compared with some of the other sources of error described next.

Since the reflectivity factor Z depends on the sixth power of the drop diameter, the Z-R relationship, unlike the attenuation-rainfall relationship, depends sensitively on the nature of the raindrop size distribution. There have been many studies of the Z-R relation and there are unfortunately, almost as many Z-R relationships as there are studies. Thus it has been common practice in the past to use a "standard" relationship, $Z = 200R^{1.6}$, based on the results of Marshall and Palmer (1940). By taking into account the character of the precipitation system, it may be possible to do better than this. The fact remains, however, that the spatial and temporal variability of drop-size distribution is so great that in practice the best approach is to use a number of raingauges to calibrate the radar.

The curvature of the earth causes the radar beam to rise a considerable distance above the surface at long ranges. This is true, in most situations, even after one has allowed for the curvature of the radar beam due to atmospheric refraction. This leads to several difficulties in the measurement of surface rainfall intensity. One problem is the non-vertical orientation of precipitation shafts owing to wind shear. This may be quite troublesome in the case of slow-falling snow particles and, according to Harrold et al. (1974), is not negligible even for raindrops falling at several meters per second. Another problem caused by measuring reflectivity aloft rather than at the surface is that it fails to take into account growth or evaporation of rain at low levels. A third problem caused by the elevation of the radar beam is the increased likelihood of the beam intercepting sleet in the melting layer. The sleet layer is seen by radar as an intense band of echo which serves to confuse the interpretation of rainfall intensity. Use of a calibration rain-gauge in an area contaminated by interception of the bright band is the best solution where such a gauge exists. In general, it is important to minimize the number of occasions when the bright band is intersected at close range by using a narrow beam at the lowest possible elevation angle. Of course the most serious problem of all is that at very long ranges the beam becomes incompletely filled by precipitation as the top of the beam begins to overshoot the precipitation layer altogether. This is usually the factor which determines the ultimate limit to the range of detectability of precipitation. However, the other difficulties mentioned above usually restrict the quantitative measurement of surface rainfall to ranges very much closer than this.

The previous discussion emphasizes the need to keep the radar beam low. By doing this, however, a fresh set of problems is incurred because, in the case of a radar sited amidst irregular terrain with the beam directed almost horizontally, some or all of the beam will be intercepted by hills at various ranges. Although one attempts to minimize these difficulties by locating the radar at an unobstructed site, these difficulties can never be avoided altogether.

It is interesting to compare the accuracy of radar with that achievable with raingauge network of different densities. The accuracy of the raingauge networks depends critically on the nature of the rain. The radar measurements, on the other hand, were nearly independent of rainfall type. The radar system calibrated using two raingauges over the 1000-square-kilometer experimental area had the same accuracy as a raingauge network of nine gauges per 100 square kilometers in the presence of typical widespread rain. In showery situations, the same radar system had an accuracy comparable with a raingauge network with a density of about gauges per 1000 square kilometers.

New Words

quantitative	[ˈkwɔntitətiv]	a. 量的,定量的
raingauge	[ˈreingedʒ]	n. 雨量计
telemetry	[tiˈlemitri]	n. 遥测法
assessment	[əˈsesmənt]	n. 评价,估定
utility	[juːˈtiliti]	n. 有用,有益
attenuation	[əˌtenjuˈeiʃən]	n. 衰减
backscatter	[ˈbækˌskætə]	n. 后向散射
reflectivity	[ˌriːflekˈtiviti]	n. 反射率
median volume diameter	[ˈmiːdjən][ˈvɔljuːm][daiˈæmitə]	中值体积直径
exploit	[iksˈplɔit]	vt. 利用
torrential	[təˈrenʃəl]	a. 猛烈的
aerial	[ˈɛəriəl]	n. 天线

orientation	[ˌɔ(:)rien'teiʃən]	n. 方位,取向
negligible	['neglidʒəbl]	a. 可以忽视的
aloft	[ə'lɔft]	ad. 在上面,在空中
elevation	[ˌeli'veiʃen]	n. 高升,仰角
likelihood	['laiklihud]	n. 可能
beamwidth	['biːmwidθ]	n. 波束宽度
simultaneously	[siməl'teiniəsli]	ad. 同步地,同时地
azimuth	['æziməθ]	n. 方位角
axis	['æksis]	n. 轴
parameter	[pə'ræmitə]	n. 参数
dielectric	[ˌdaii'lektrik]	n. 电介质
summation	[sʌ'meiʃən]	n. 总数
valid	['vælid]	a. 正确的
fluctuation	[ˌflʌktju'eiʃən]	n. 变动,起伏,扰动
random	['rændəm]	n. 随机
clutter	['klʌtə]	n. 地物杂波
discrimination	[disˌkrimi'neiʃən]	n. 辨别,区别
calibrate	['kælibreit]	vt. 校准
sleet	[sliːt]	n. 冻雨,雨夹雪
contaminate	[kən'tæmineit]	vt. 污染
intersect	[ˌintə'sekt]	v. 相交
amidst	[ə'midst]	prep. 在…中间
density	['densiti]	n. 密度,稠密

Supplemental Reading 20.1

The leading-line/trailing-stratiform type of MCS typically exhibits "rear inflow," which is a layer of low θ_e air that enters the MCS from the rear below the trailing anvil cloud of the stratiform region and descends toward the leading convective line (Fig. 1). The descent is gradual across the stratiform region but often takes a sudden plunge downward as it approaches the immediate rear of a region of active convective cells.

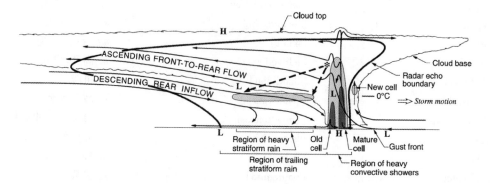

Fig. 1 Conceptual model of the kinematic, microphysical, and radar echo structure of a convective line with trailing-stratiform precipitation viewed in a vertical cross section oriented perpendicular to the convective line (and generally parallel to its motion). Intermediate and strong radar reflectivity is indicated by medium and dark shading, respectively. H and L indicate centers of positive and negative pressure perturbations, respectively. Dashed-line arrows indicate fallout trajectories of ice particles passing through the melting layer. From Houze et al (1989)

Supplemental Reading 20.2

Role of MCSs in Tropical Cyclone Development

Satellite data show that tropical cyclones spin up from MCSs. It appears that the MCVs in the stratiform regions of the MCSs are the origin of tropical cyclone circulations. It has been suggested that the middle level vortex in the stratiform region evolves into a deep tropical cyclone circulation (Velasco and Fritsch, 1987; Miller and Fritsch, 1991; Fritsch et al., 1994; Fritsch and Forbes, 2001). Bister and Emanuel (1997) suggested that cooling below the base of the stratiform cloud (of the type discussed by Zhang (1992)) is involved in the extension of the middle level vortex downward in tropical MCSs. They proposed that when the cooling-induced MCV extended low enough, it could connect with the boundary layer and develop into a tropical cyclone. The mechanism by which the developing cyclone builds downward and connects with the surface layer, however, remains unclear. Ritchie and Holland (1997), Simpson et al (1997), and Ritchie et al (2003) hypothesize that the primary hurricane vortex forms and builds downward when two or more MCSs interact. According to this idea each MCS spins up its own MCV in the stratiform region of the MCS as a result of the profile of heating aloft and cooling at lower levels (Fig. 2). When two or more MCVs are in close proximity, they begin to rotate around a common axis and amalgamate into a common vortex.

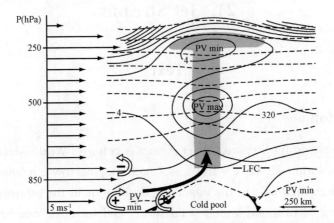

Fig. 2 Conceptual diagram of the structure and redevelopment mechanism of a mesoscale warm core vortex associated with an MCS. Thin arrows along the ordinate indicate the vertical profile of the environmental wind. Open arrows with plus or minus signs indicate the sense of the vorticity component perpendicular to the plane of the cross section produced by the cold pool and by the environmental vertical wind shear. The bold solid arrow indicates the updraft axis created by the vorticity distribution. Frontal symbols indicate outflow boundaries. Dashed lines are potential temperature (5 K intervals), and solid lines are potential vorticity (2×10^{-7} $m^2 \cdot s^{-1} \cdot K \cdot kg^{-1}$ intervals). The system is propagating left to right at about $5-8$ m s^{-1} and is being overtaken by air of high equivalent potential temperature in the low level jet. Air overtaking the vortex ascends isentropic surfaces, reaches its level of free convection (LFC), and thereby initiates deep convection. Shading indicates cloud. From Fritsch et al (1994)

21. Jet Streams

Text

(1) Definition

(a) <u>Atmospheric Jet Streams.</u> A jet stream is defined [World Meteorological Organization, Resolution 25 (EC — IX)] as "A strong narrow current, concentrated along a quasi-horizontal axis in the upper troposphere or in the stratosphere, characterized by strong vertical and lateral wind shears and featuring one or more velocity maxima." In addition, the following characteristic criteria are recommended: "Normally a jet stream is thousands of kilometers in length, hundreds of kilometers in width and some kilometers in depth. The vertical shear of wind is of the order of 5—10 m/sec per km and the lateral shear is of the order of 5 m/sec per 100 km. An arbitrary lower limit of 30 m/sec is assigned to the speed of the wind along the axis of a jet stream."

(b) <u>Jet Streams in General.</u> The name "jet stream" refers mainly to atmospheric phenomena meeting the above qualifications. Ocean currents, such as the Gulf Stream, Kuroshio, etc., follow the same physical laws and could be considered as jet steams, although their velocities are considerably smaller than atmospheric examples. Jet stream-like flow patterns may also be generated in laboratory experiments in which water tanks ("dishpan") are rotated (Fig. 1). These tanks may be heated along the

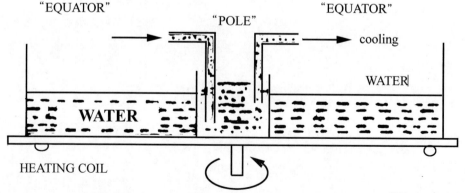

Fig. 1 Cross section through rotating "dishpan"

outer rim ("equator") and cooled at the center ("pole"). This simulating atmospheric conditions. The basic physical principles of jet stream formation may be studied from such experiments.

(2) Formation of Jet Streams

(a) <u>Conservation of Absolute Angular Momentum.</u> Heat input by absorption of solar radiation, mainly at the ground at low latitudes, and heat loss by excessive radiation into space at high latitudes necessitate a heat transport from equator to pole. If the earth did not rotate, this would result in a simple "trade wind" or "Hadley" cell, as shown in Fig. 2.

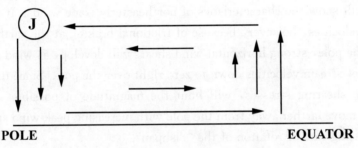

Fig. 2 "Trade wind" cell on nonrotating earth

In a rotating system, such as the earth's atmosphere or the oceans, in addition to heat, absolute angular momentum G (measured with respect to an absolute coordinate system which does not rotate with the earth) will be transported with the meridional circulation generated by differential heating and its resultant pressure forces. If no external forces were acting on an air mass, its G would be conserved.

$$G = (\Omega + \omega_1)(R\cos\psi_1)^2 = (\Omega + \omega_2)(R\cos\psi_2)^2 \tag{1}$$

where Ω is the angular velocity of the earth (7.292116×10^{-5} rad/sec), ω is the west-east component of angular velocity of air relative to the earth's surface, R is the radius of earth (about 6370 km) and ψ is the geographic latitude. Subscripts "1" and "2" indicate conditions at the beginning and the end of meridional displacement of air. Considering that zonal (west to east) wind speed is given as $V = \omega R\cos\psi$, we may conclude that air displaced from 30°N to 40°N would acquire a westerly velocity of 99 m/sec by conserving its absolute angular momentum during the displacement.

According to Fig. 2, air rising over the equator has been in frictional contact with the earth's surface, acquiring it a angular momentum which is characteristic for these latitudes. As the air travels poleward underneath the tropopause it will——under conservation of angular momentum——turn from a southerly into a westerly wind direction. It will, at the same time, increase its westerly velocity with the distance over which it moves meridionally. As seen from the above equation and the numerical example, relatively small meridional displacements suffice to generate a strong westerly jet stream.

(b) <u>Jet Streams in Laboratory Experiments.</u> In a slowly rotating system (as may be simulated in "dishpan" experiments), the meridional circulation may extend close to the pole, generating a jet stream at high latitudes (indicated by "J" in Fig. 2). This is possible, because with slow rotation, Ω in Eq. (1) is small, hence only small amounts of absolute angular momentum need to be transported. The equatorward return flow will show the characteristics of northeasterly trade winds. It will not reach jet-stream velocities, however, because of frictional braking at the earth's surface.

Near the pole, strong horizontal wind shears will develop, as wind speeds should drop from jet-stream velocities down to zero right over the pole. Since turbulence, resulting from shearing stresses, will limit the magnitude of possible shear, the jet stream will move farther away from the pole with increasing peak wind speeds, caused by an increasing rate of rotation of the "dishpan".

Beyond a critical rate of rotation, the meridional circulation shown in Fig. 2 is not capable any longer of transporting heat and angular momentum in the required amount. The jet stream breaks down into a number of symmetric waves or "eddied". Figure 3 shows schematically a symmetric three-wave arrangement in a rotating dishpan. Such waves were first described theoretically by Rossby. With still increasing rotation rates, the number of such waves increases. Up to seven of such symmetric waves have been observed in laboratory experiments.

Beyond that, irregular waves, very similar to the ones observed on daily hemispheric weather maps, will form. These waves have a strong "tilt" of their axes (Fig. 4), thus permitting more westerly momentum and heat to flow northward on the east side of troughs than flows southward on the west side of troughs.

Northward transport of zonal momentum, which is equivalent to angular momentum, may be considered in terms of the product $\rho \cdot \overline{u'v'}$, where ρ is the air density and u', v' are the zonal (west to east) and meridional (south to north) wind components in terms of departures from mean conditions around the latitude circle of observation; the "bar" indicates averaging over all longitudes of the hemisphere. As may be seen from Fig. 4, strong components of $u' > 0$ are multiplied with $v' > 0$ east of

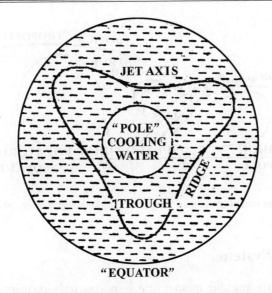

Fig. 3 Three symmetric Rossby waves in rotating dishpan

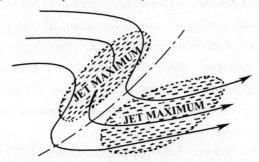

Fig. 4 Streamlines and jet streams in a "tilting" low-pressure trough

the trough, while west of the trough, light east wind components ($u' < 0$) are multiplied with strong north winds ($v' < 0$). In both instances, $u'v' > 0$, indicating northward transport of zonal momentum.

(c) <u>Atmospheric Conditions.</u> Conditions of rotation and heat transfer in the earth's atmosphere are such that irregular waves of the characteristics described above will form. Because of the large momentum transports resulting from the relatively fast rate of rotation of the earth, the trade wind cell of Fig. 2 cannot extend close to the pole. Enormous jet-stream speeds would result from Eq. (1), causing impossibly high wind speeds. Instead the trade wind cell is——on the average——confined to latitudes equatorward of 30°. The circulation model shown in Fig. 5, and described by Palmen is observed, revealing more than just one jet stream.

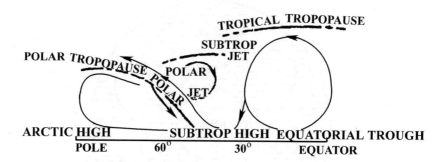

Fig. 5 Schematic diagram of mean meridional circulation during winter, Northern Hemisphere (after Palmen)

(3) Jet Stream Systems

Although absolute angular momentum is not strictly conserved in the motion of air masks because of the action of external pressure forces, at least a tendency toward such conservation should lead to jet-stream formation in regions where considerable meridional displacement of air masses takes place. Such displacements usually reach extreme values at characteristic atmospheric interfaces, prescribed by the thermal structure of the atmosphere, and designated as "opuses" (Fig. 6). We may, therefore, roughly classify jet streams into tropopause, stratopause and mesopause jet streams.

(a) <u>Tropopause Jet Stream. Subtropical Jet Stream (STJ).</u> This is found near 30°N during Northern Hemisphere winter as a continuous belt around the hemisphere, with three waves superimposed. Wave crests, with jet maxima (i. e. , areas of maximum speed along the axis of strongest flow, the latter called the "jet axis") are located near 70°W, 40°E and 150°E, the latter maximum showing the strongest mean winds (e. g. , >150 knots for a winter season 1955—1956). The maximum winds ever measured by balloon (150 m/sec=291 knots) were reported in this jet maximum by Arakawa. According to Fig. 5, the STJ draws its high speeds from the circulation associated with the Hadley or trade wind cell. Part of the momentum transport within this circulation is accomplished by a mean circulation in a meridional plane and part by horizontal eddy transport, the latter mainly within the three waves mentioned above. As shown by Palmen, Alaka and others, the eddy transport gains in relative importance over the meridional-circulation transport with increasing altitude.

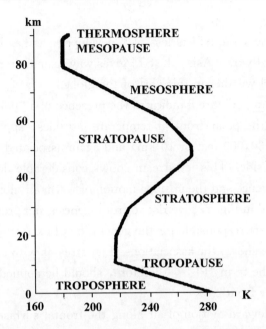

Fig. 6 Thermal structure and nomenclature of the standard atmosphere

The core of highest wind speeds in the STJ is found near the 200-hPa level (about 40,000 feet or 12 km). The tropopause shows a discontinuity in height (i. e. , tropopause "gap" or "break") on its poleward side. A "baroclinic zone" (i. e. , a region with quasi-horizontal temperature gradients along its baric surfaces extends underneath the STJ but usually does not extend below the 400-millibar level. The lower troposphere underneath the STJ is generally "barotropic", meaning that horizontal and vertical wind shears as well as horizontal temperature gradients are small. This is in accordance with the subsidence found on the average underneath the STJ (Fig. 5) which, together with the divergent flow at the earth's surface would tend to destroy any fronts or "baroclinic zones".

From the circulation model shown in Fig. 5 and from the above discussion, we may infer that the STJ is located approximately at tropopause level over the subtropical high-pressure ridge. Since in the Southern Hemisphere winter this high-pressure ridge is closer to the equator than it is in the Northern Hemisphere, we may deduce the same for the southern STJ. Conditions in the Southern Hemisphere have not yet been studied extensively, and there is a lack of data. Analyses by Gibbs and Hutchings over Australia and the Pacific confirm the above conclusions. Strong westerlies, at times, have been observed near the tropopause over Recife (Brasil), approximately

9° from the equator.

During the summer, the STJ is less well developed, as has been pointed out by Riehl (1962). Especially over Asia, the STJ gives way to an easterly jet stream, associated with monsoonal weather in this part of the globe.

Frontal Jet Streams. Figure 5 indicates the presence of a "polar-front jet stream" (PFJ), located over the polar front of temperate latitudes, approximately near the 300-hPa level (about 30,000 feet or 10 km) and again associated with a "tropopause gap" on its poleward side. This jet stream shows considerable day-to-day variation. The low-level convergence and the upslope motion of warm air along the frontal zone, schematically indicated in Fig. 5, give rise to frontogenesis and precipitation in this region. The PFJ is in part responsible for the generation of extratropical cyclones. Galloway and others have shown the associations of jet streams with continental and maritime arctic fronts. The term "PFJ", therefore, should be applied in a broader sense than just to the "polar front".

Although the poleward motion of air along the frontal surface, shown in Fig. 5, would support our previous conclusion that conservation of absolute angular momentum is responsible for jet stream formation, atmospheric conditions near the PFJ are more complex than would appear from such a simplifying statement. Individual jet maxima (region of strongest winds) are frequently found at the tips of deep low-pressure troughs, rather than in the crest of high-pressure ridges, as the angular momentum theory would suggest and as the STJ bears out. We may conclude, therefore that pressure forces, and the release of potential energy by sinking of cold air and rising of warm air in the vicinity of frontal surfaces, have a controlling influence on the formation and maintenance of the PFJ.

(b) Stratopause Jet Streams. The stratopause located near the 50-km level harbors a strong westerly jet stream (Stratospheric Jet) during the winter season, with winds frequently in excess of 150 knots near 70°N as shown by Godson, Lee and others. A two-wave pattern dominates, with one trough over eastern Siberia and another over Hudson Bay, suggesting a strong orographic influence of the Himalayas and the Rocky Mountains even at these latitudes. Large-scale horizontal and vertical transport processes within this jet stream have to be held responsible for observed seasonal ozone concentration anomalies as reported for instance by Goetz, which cannot be explained from the radiation budget at these latitudes. Transport of radioactive debris in the stratosphere is also strongly influenced by this jet stream.

With continuous insolation over the summer pole, the horizontal temperature and pressure gradients in the stratosphere reverse from winter to summer. Consequently, the winter jet stream breaks down and gives way to weaker easterlies during summer. The stratospheric flow patterns in the Southern Hemisphere show similar seasonal trends. The summer easterlies seem to be weaker there, however, than in the northern hemisphere.

The breakdown of the winter circulation pattern may be rather abrupt. While in the Southern Hemisphere it falls close to the period of equinoxes, in the northern hemisphere large variations in the breakdown period may occur from one year to another. As the orientation and intensity of the stratospheric jet changes, stratospheric temperatures over stations underneath this let stream may experience "explosive warming" of more than 20℃ within a few days.

(c) <u>Mesopause Jet Streams.</u> These are still poorly explored. Because they occur in the ionosphere and consist of drifts of charged-particle cloud, they are sometimes referred to as electrojets. Electromagnetic forces have to be taken into account in considering air motions at these altitudes (>80 km).

(4) Effects of Jet Streams

For motions in an isobaric surface the quantity

$$\zeta_a = \frac{V}{r} - \frac{\delta V}{\delta n} + f$$

called the vertical component of the absolute vorticity, obeys the following equation

$$\frac{d\zeta_a}{dt} = -D\zeta_a \tag{2}$$

where V is the velocity of air current, r is the radius of curvature of streamline, n is the coordinate normal to the flow direction, $f = 2\Omega \sin \psi =$ Coriolis parameter (i. e., vorticity component of rotating earth normal to earth's surface), and D is the divergence of flow.

The strong shears on either side of a jet maximum (together with possible effects of curvature of flow) generate positive (cyclonic) vorticity poleward and negative (anticyclonic) vorticity equatorward of the axis of a westerly jet. Air flowing through this vorticity pattern will undergo divergence or convergence according to Eq. (2)——divergence ($D>0$) if the vorticity of an air parcel decreases along its way ($\frac{d\zeta_a}{dt} < 0$), and convergence if the vorticity increases. Figure 7 shows schematically the resulting divergence and convergence pattern around a jet maximum, superimposed upon straight flow. Curvature will have a modifying influence on this pattern.

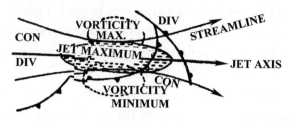

Fig. 7　Schematic position of convergence and divergence at jet stream level (near tropopause), and of fronts at earth's surface

Since divergence at tropopause level will result in a mass deficit in the total vertical air column, surface pressure falls will result underneath the divergence region. If a frontal zone at low levels is present at the same time, cyclogenesis may occur, as has been shown by Petterssen, Newton and others. Thus, weather patterns, specifically cloudiness, precipitation and cyclones are intimately linked to jet streams at tropopause level.

Strong winds in the jet-stream region have an important effect on airline operations and on the planning of their economy. Records of flying time have been achieved in transcontinental and transoceanic flights by "riding" the jet stream. Conversely, headwind flights try to avoid the jet stream whenever possible under Air Traffic Control regulations. "Minimum Flight Path Planning" takes into account the effect of winds, specifically of jet streams, on flying time.

Strong shears in the jet-stream region may cause clear air turbulence, i.e., bumpy flight conditions that may, under serious conditions, lead to material fatigue and even to aircraft failure, not to speak of passenger discomfort.

New Words

lateral	['lætərəl]	a.	横向的,侧向的
recommend	[rekə'mend]	v.	建议;推荐
arbitrary	['aːbitrəri]	a.	任意的,任定的
assign	[ə'sain]	v.	分配,指定
qualification	[kwɔlifi'keiʃən]	n.	条件;限制;资格
tank	[tæŋk]	n.	箱,槽,坦克
dishpan	['diʃpæn]	n.	平底锅
simulate	['simjuleit]	v.	模拟
turntable	['təːnteibl]	n.	转台

necessitate	[niˈsesiteit]	v. 使有必要
meridional	[məˈridiənl]	a. 经向的,子午的
resultant	[riˈzʌltənt]	a. 合成的,作为结果而产生的 n. 合力,合矢量
radian (rad.)	[ˈreidiən]	n. 弧度
zonal	[ˈzəunl]	a. 纬向的
acquire	[əˈkwaiə]	v. 取得;达到
underneath	[ʌndəˈniːθ]	prep. 在…下面
brake	[breik]	n. v. 制动
peak	[piːk]	n. 山峰;峰值,最大值
schematically	[skiˈmætikəli]	ad. 概要地,图示方法
arrangement	[əˈreindʒmənt]	n. 排列,分布;安排
irregular	[iˈregjulə]	a. 不规则的
tilt	[tilt]	v. n. 倾斜
trough	[trɔf]	n. 槽;波谷
longitude	[ˈlɔndʒitjuːd]	n. 经度
multiply	[ˈmʌltiplai]	v. 乘
reveal	[riˈviːl]	v. 揭示,显示
cross section	[krɔs][ˈsekʃən]	截面
Gulf Stream	[gʌlf][striːm]	(墨西哥)湾流
Kuroshio	[kuˈrəuʃiːəu]	(日本)黑潮
Hardley cell	[ˈhɑːdli][sel]	哈得来环流
prescribe	[priˈskraib]	v. 规定,指示
nomenclature	[nauˈmənklətʃə]	n. 名称,术语
subtropical	[ˈsʌbˈtrɔpikəl]	a. 亚热带的,副热带的
superimpose	[ˈsjuːpərimˈpəuz]	v. 增添;迭加
baroclinic	[bærəˈklinik]	a. 斜压的
barotropic	[bærəˈtrɔpik]	a. 正压的
subsidence	[səbˈsaidəns]	n. 下沉;减小
destroy	[disˈtrɔi]	v. 破坏,消灭
ridge	[ridʒ]	n. 脊,波峰
deduce	[diˈdjuːs]	v. 推断,推演
confirm	[kənˈfəːm]	v. 确定,证实
monsoonal	[mɔnˈsuːnəl]	a. 季风的
frontogenesis	[ˈfrʌntˈdʒenisis]	n. 锋生
extratropical	[ˈekstrəˈtrɔpikəl]	a. 热带以外的,温带的
continental	[kɔntiˈnentl]	a. 大陆的,大陆性的
maritime	[ˈmæritaim]	a. 海洋性的,沿海的
arctic	[ˈɑːktik]	a. 北极圈的,北冰洋的
tip	[tip]	n. 尖端,末端
maintenance	[ˈmeintinəns]	n. 保持,维持
dominate	[ˈdɔmineit]	v. 统治,支配
Siberia	[saiˈbiəriə]	n. 西伯利亚
bay	[bei]	n. 海湾

debris	[ˈdebri]	n. 碎片
trend	[trend]	n. v. 趋势,倾向
abrupt	[əbˈrʌpt]	a. 突然的
orientation	[ɔːrienˈteiʃn]	n. 方向,走向
explosive	[iksˈpləusiv]	a. 爆炸性的
mesopause	[ˈmesəpɔːz]	n. 中层顶
explore	[iksˈplɔː]	v. 探索,探测;研究
electrojet	[iˈlektrəudʒet]	n. 电急流
curvature	[ˈkəːvətʃə]	n. 曲率、弯曲
streamline	[ˈstriːmˈlain]	n. 流线
vorticity	[vɔːˈtisiti]	n. 涡度
undergo	[ʌndəˈgəu]	v. 经受
cyclogenesis	[ˈsaikləuˈdʒenisis]	n. 气旋生成
airline	[ˈɛəlain]	n. 航线;航空公司
transcontinental	[ˈtrænzkɔntiˈnentl]	a. 横贯大陆的
flight	[flait]	n. 飞行
conversely	[ˈkɔnvəːsli]	ad. 相反
traffic	[ˈtræfik]	n. 交通
bumpy	[ˈbʌmpi]	a. 颠簸的
fatigue	[fəˈtiːg]	n. 疲劳
passenger	[ˈpæsindʒə]	n. 旅客
discomfort	[disˈkʌmfət]	n. 不舒服
in accordance with	[əˈkɔːdəns]	与……一致
give way to	[giv][wei]	让位于,为……代替
bear out	[bɛə]	证明,证实
in excess of	[ikˈses; ˈekses]	超过
head-wind	[hed-waind]	迎风的,顶头风
traffic regulation	[ˈtræfik][regjuˈleiʃən]	交通规则
material fatigue	[məˈtiəriəl][fəˈtiːg]	材料疲劳

22. Meteorology

Text

Meteorology is the science of the atmosphere, including the study of weather. Most weather originates within the troposphere, the atmospheric layer about 8 mile (13 km) thick that immediately overlies the earth's surface and contains 90% of the atmosphere by weight. Conditions in the troposphere change rapidly; for this reason, weather predictions beyond a two-week period remain unreliable despite technological advances.

Except for radiation, which can be observed by weather satellites, the processes that produce and alter the weather cannot be measured directly but must be calculated from observations of atmospheric variables. The principal variables measured by meteorologists are temperature, precipitation (rain or snow), wind, humidity, cloudiness, air pressure, and air pollutants.

Development of Meteorology

Before the late 19th century, when balloons first reached altitudes of 10 mile (16 km), weather observers had to rely mainly on what they could learn from the ground. Most of this information was qualitative. Aristotle's great treatise Meteorological, written about 350 B. C., is the outstanding work of this era. It was not until some 2,000 years after Aristotle, after Galileo's invention of the thermometer about 1593 and the discovery of the barometric principle by Torricelli in 1643, that the first measurements were taken and records kept. The longest continuous series of observations has been recorded at Paris since 1664; the longest series in the United States has been maintained at New Haven, Conn., since 1779.

Comparisons of weather observations from different locations led to the concept of moving weather systems. In 1743 Benjamin Franklin used weather reports gathered by mail to trace the courses of severe storms. He found that many storms arrived later in Boston than in Philadelphia, although the winds along the Atlantic coast were blowing from the northeast. Networks of observing stations and, in the early 19th century, the invention of the telegraph allowed weather charts to be prepared from observations taken at the same time over large areas. Soon it became evident that air moves in large clockwise and counterclockwise whirls that cover a circular area 500—1000 mi (805—1609 km) across. These are called anticyclones and cyclones, respectively, in the Northern Hemisphere; in the Southern Hemisphere they rotate in the opposite direction. Between latitudes 30° and 60° they generally move toward the east, traveling 500—1000 mi (805—1609 km) per day and carrying their cloud systems with them.

    Nineteenth-century observers learned that anticyclones are usually fair-weather areas, and that storminess, precipitation, and rapid temperature changes occur in cyclones. This picture was considerably sharpened by the Norwegian meteorologist Vilhelm Bjerknes and his son Jakob. In 1920 they found that changing temperatures and bad weather occurred mainly in association with sharp lines of wind change, which they labeled "fronts". Along the warm front ahead of a cyclone center, warm air arrives from tropical regions. At the cold front this warm air mass is replaced by a fresh outbreak of cold air from the polar zone. This discovery gave forecasters a model with which to analyze weather phenomena. If weather changes occurred according to a logical pattern, predictions could be made using mathematical calculations.

    Meteorology entered a period of rapid advance about the time the Bjerknes' cyclone model appeared. Measurements in the upper air became routine, spurred by advances in aviation. The airplane itself provided a means of measuring pressure, temperature, and humidity at higher and higher altitudes. Winds were studied by observing the paths taken by balloons released from land stations. In the 1930s came the development of the radiosonde, an instrument that could be attached to weather balloons to transmit information on pressure, temperature, and humidity during its ascent. Since the perfection of radar in the 1940s the radiosonde balloons have been tracked with radio signals, permitting wind determinations to be made even when the skies are overcast.

    Technological developments since World War II have expanded man's knowledge of the atmosphere. Today, information is collected by aircraft and ocean-going vessels, by drifting and moored buoys, and by landbased weather stations. Radar tracking systems measure turbulence, wind velocity, air pollution, and composition of the atmosphere. Weather surveillance satellites constantly monitor the globe, enabling meteorologists to detect a new weather system from its inception. Computers assess the collected data and perform mathematical calculations that project weather conditions for days or weeks ahead. International cooperation on global weather study has greatly increased the potential for extended forecasting.

Weather Processes

Technological advances have allowed meteorologists to make more accurate studies of the processes that determine the weather. Some of the more important subjects of these studies are discussed hereafter.

Radiation is the process by which energy, in the form of electromagnetic waves, is transmitted from the sun to the earth, the atmosphere, and back to space. Virtually all meteorological phenomena occur as a result of radiation. About two thirds of incoming solar energy are absorbed by the earth's surface and by water vapor and carbon dioxide in the atmosphere. The remaining third is reflected back to space by the earth, the atmosphere, and clouds. The resulting gain in heat by the earth is balanced by heat loss, particularly through the evaporation of water, a process that requires the expenditure of energy.

Radiation received by the earth is unequally distributed, as is the energy transferred from earth to atmosphere. Ultimately, all air motions and weather systems are caused by the heat flow from warm to cold regions that is prompted by this unequal heating. In particular, the polar regions constantly lose heat during winter, while the tropics gain energy. Their climates are modified by great wind systems that constantly move warmer air poleward and colder air toward the equator.

With various types of satellites solar radiation and its effects can be observed and measured on a worldwide basis.

Turbulent motions (turbulence) are random, small-scale motions that are responsible for transferring heat, moisture, and other matter to the atmosphere. They also play an important role in the dissipation of energy, as it is passed from larger-to smaller-scale motions and thence to thermal energy, or heat.

Turbulence occurs when wind speeds fluctuate and when heating of the earth's surface creates buoyancy. When winds are not active and turbulence is inhibited, fumes and other pollutants remain near the earth's surface as smog. Turbulent motions also disperse plant seeds, viruses, and other organisms throughout the troposphere.

Because of their erratic nature, turbulent motions are best analyzed statistically. Computer modeling of atmospheric turbulence near the ground is used to calculate the dispersion of pollutants and to identify those conditions that lead to excessive concentrations of pollutants.

Clouds are accumulations of water droplets or ice crystals. Condensation of water vapor takes place on tiny particles of salt, dust, or smoke, called condensation nuclei that exist in abundance in the air. Small water droplets form when the air is near 100% relative humidity, even at temperatures far below freezing. To grow to raindrop size, the cloud droplets must increase in diameter by as much as 100 times.

Ice crystals develop on freezing nuclei that originate as dust from certain soils, or possibly as meteoric dust. Since freezing nuclei are far less numerous than condensation nuclei, water droplets may persist without freezing at temperatures as low as $-40°F$ ($-40°C$), a condition known as supercooling. If ice-crystals enter a supercooled cloud, it may become an ice-crystal cloud. We can convert supercooled clouds to ice-crystal clouds by introducing artificial nuclei, such as silver iodide, to the atmosphere.

The interactions of ice crystals and supercooled droplets influence the electrical charge of a cloud. As small droplets freeze in convective clouds, electricity is generated. The charges separate, the positive charges going to the top of the cloud and the negative charges to the bottom. When the voltage difference between them is great enough, lightning occurs.

General Air Circulation

Weather occurs on many scales. The largest air movement, called general circulation, includes the wind systems that give rise to the changing daily weather. In turn, these wind systems control small-scale motions such as thunderstorms. The constant circulation of air in the atmosphere causes the vast differences in temperature, precipitation, wind and cloudiness that take place over the earth in a given year.

As Carl-Gustaf Rossby discovered in 1940, the broad upper-air currents blowing from the west in the temperate zone form a chain of "long waves" so called because the east-west length of one wave may be as great as 3000—5000 mi (4826—846 km). The number of long waves, their rate of motion, and their amplitude all vary with time, and they largely control the changing weather patterns. Within this wave like westerly current lies a central band of winds, called the jet stream, with velocities of 150—200 mph (241—322 km/h) and occasionally more. The greatest weather changes usually are found along the axis of this band.

Coriolis Force

The flow of warm air poleward and cool air toward the equator which balances the distribution of the earth's heat, is the impetus behind all air motions. To understand the general circulation of the air, however, one must also consider the rotation of the earth. In general, winds are described from the view-point of a stationary observer on earth. Since this frame of reference—the earth—is rotating, motions that would be straight in a stationary frame of reference appear curved to the observer on earth. This effect is called the Coriolis force, after the French mathematician GaSpard Gustave Coriolis. The winds follow this pattern: they are deflected to the right in the Northern Hemisphere and to the left in Southern Hemisphere. The earth's rotation, therefore, produces winds that move to the east and west as well as to the north and south.

Large-scale atmospheric motions exist mainly because of the earth's rotation. In particular, air tends to move at constant speed, balanced by the interaction of air pressure and the Coriolis force. Such a balanced wind is referred to as a geostrophic wind. As a consequence, air motion largely follows the lines of constant pressure, or isobars. Because of this important relationship, the analysis of pressure distribution in the atmosphere is an important meteorological tool.

New Words

originate	[əˈridʒneit]	v. 起源
overlie	[əuvəˈlai]	v. 位于上部
Aristotle	[ˈæristɔtl]	n. 亚里士多德
treatise	[ˈtriːtiz]	n. 论文,论著
Conn. (Connecticut)	[kəˈnetikət]	n. [美国]康涅狄格州
whirl	[wəːl]	n. 涡旋
spur	[spəː]	v. 鼓舞,推动
radiosonde	[ˈreidiəusɔnd]	n. 无线电探空仪
drift	[drift]	v. n. 漂移
moor	[muə]	v. 使停泊
buoy	[bɔi]	n. 浮标
landbased	[lændbeist]	a. 陆基的
surveillance	[səːˈveiləns]	n. 监视
monitor	[ˈmɔnitə]	v. n. 监视,监测

inception	[in'sepʃn]	n. 开始
hereafter	[hiər'a:ftə]	ad. 今后
random	['rændəm]	a. 偶然的,随机的
thence	[ðens]	ad. 因此,从此
fluctuate	['flʌktjueit]	v. 摆动,波动
buoyance	['bɔiəns]	n. 浮力
inhibit	[in'hibit]	v. 抑制
fume	[fju:m]	n. (气味强烈的)烟、汽
virus	['vaiərəs]	n. 毒素,病毒
erratic	[i'rætik]	a. 无规则的
meteoric	[mi:ti'ɔrik]	a. 陨石的
iodide	['aiəudaid]	n. 碘化物
impetus	['impitəs]	n. 推力,刺激
frame	[freim]	n. 框架,结构
except for	[ik'sept]	除…之外
in abundance	[ə'bʌndəns]	丰富地,大量地
frame of reference	[freim]['refrəns]	参考系

23. Numerical Weather Prediction

Numerical weather prediction is the prediction of weather phenomena by the numerical solutions of the equations governing the motion and changes of conditions of the atmosphere. More generally, the term applies to any numerical solution or analysis of the atmospheric equations of motion.

The laws of motion of the atmosphere may be expressed as a set of partial differential equations relating the instantaneous rates of change of the meteorological variables to their instantaneous distribution in space. These are developed in dynamic meteorology. A prediction for a finite time interval is obtained by summing the succession of infinitesimal time changes of the meteorological variables, each of which is determined by their distribution at the preceding instant of time. Although this process of integration may be carried out in principle, the nonlinearity of the equations and the complexity and multiplicity of the data make it impossible in practice. Instead, one must resort to finite-difference approximation techniques in which successive changes in the variables are calculated for small, but finite, time intervals at a finite grid of points spanning part or all of the atmosphere. Even so, the amount of computation is vast, and numerical weather prediction remained only a dream until the advent of the modern high-speed electronic computing machine. These machines are capable of performing the millions of arithmetic operations involved with a minimum of human labor and in an economically feasible time span. Numerical methods are gradually replacing the earlier, more subjective methods of weather prediction in many United States government weather services. This is particularly true in the preparation of prognoses for large areas. The detailed prediction of local weather phenomena has not yet benefited greatly from the use of numericodynamic methods.

Short-Range Numerical Prediction

By the nature of numerical weather prediction, its accuracy depends on (1) an understanding of the laws of atmospheric behavior, (2) the ability to measure the instantaneous state of the atmosphere, and (3) the accuracy with which the solutions of the continuous equations of motion are approximated by finited-difference means. The greatest success has been achieved in predicting the motion of the large-scale (>1000 mi) pressure systems in the atmosphere for relatively short periods of time

(1—3 days). For such space and time scales, the poorly understood energy sources and frictional dissipative forces may be largely ignored, and rather coarse space grids may be used.

The large-scale motions are characterized by their properties of being quasi-static, quasi-geostrophic, and horizontally quasi-nondivergent.

These properties may be used to simplify the equations of motion by filtering out the motions which have little meteorological importance, such as sound and gravity waves. The resulting equations then become, in some cases, more amenable to numerical treatment.

A simple illustration of the methods employed for numerical weather prediction is given by the following example. Consider a homogeneous, incompressible, frictionless fluid moving over a rotating, gravitating plane in such a manner that the horizontal velocity does not vary with height. For quasi-static flow the equations of motion are Eqs. (1)

$$\frac{\partial u}{\partial t} + u\frac{\partial u}{\partial x} + v\frac{\partial u}{\partial y} = -g\frac{\partial h}{\partial x} + 2\omega v$$
$$\frac{\partial v}{\partial t} + u\frac{\partial v}{\partial x} + v\frac{\partial v}{\partial y} = -g\frac{\partial h}{\partial y} - 2\omega u \quad (1)$$

and the equation of mass conservation is Eq. (2)

$$\frac{\partial h}{\partial t} + u\frac{\partial h}{\partial x} + v\frac{\partial h}{\partial y} = -h\left(\frac{\partial u}{\partial x} + \frac{\partial v}{\partial y}\right) \quad (2)$$

where u and v are the velocity components in the directions of horizontal rectangular coordinates x and y, t is the time, g is the acceleration of gravity, ω is the angular speed of rotation, and h is the height of the free surface of the fluid. Let the variables u, v, and h be defined at the points $x = i\Delta x$, $y = j\Delta y$ ($i = 0, 1, 2, \cdots, I; j = 0, 1, 2, \cdots, J$) and at the times $t = k\Delta t$ ($k = 0, 1, 2, \cdots, K$), and denote quantities at these points and times by the subscripts i, j and k. Derivatives such as $\frac{\partial u}{\partial t}$ and $\frac{\partial u}{\partial x}$ may be approximated by the central difference quotients given by Eqs. (3)

$$\frac{\Delta_k u_{i,j}}{2\Delta t} \equiv \frac{u_{i,j,k+1} - u_{i,j,k-1}}{2\Delta t}$$
$$\frac{\Delta_i u_{j,k}}{2\Delta x} \equiv \frac{u_{i+1,j,k} - u_{i-1,j,k}}{2\Delta x} \quad (3)$$

In this way Eqs. (4), the finite-difference analogs of the continuous equations, are obtained.

$$u_{i,j,k+1} = u_{i,i,k-1} - \frac{\Delta t}{\Delta s}(u_{i,j,k}\Delta_i u_{j,k} + v_{i,j,k}\Delta_j u_{i,k} + g\Delta_i h_{j,k}) + 4\omega v_{i,j,k}\Delta t$$

$$v_{i,j,k+1} = v_{i,i,k-1} - \frac{\Delta t}{\Delta s}(u_{i,j,k}\Delta_i v_{j,k} + v_{i,j,k}\Delta_j v_{i,k} + g\Delta_j h_{j,k}) - 4\omega u_{i,j,k}\Delta t \qquad (4)$$

$$h_{i,j,k+1} = h_{i,i,k-1} - \frac{\Delta t}{\Delta s}[u_{i,j,k}\Delta_i h_{j,k} + v_{i,j,k}\Delta_j h_{i,k} + h_{i,j,k}(\Delta_i u_{j,k} + \Delta_j v_{i,k})]$$

Equations (4) give u, v, and h at the time $(k+1)\Delta t$ in terms of u, v, and h at the times $k\Delta t$ and $(k-1)\Delta t$. It is then possible to calculate u, v, and h at any time by iterative application of the above equations.

It may be shown, however, that the solution of the finite-difference equations will not converge to the solution of the continuous equations unless the criterion $\Delta s/\Delta t > c\sqrt{2}$ is satisfied, where c is the maximum value of the speed of long gravity waves $\sqrt{gh}$. Under circumstances comparable to those in the atmosphere, Δt is found to be so small that a 24-hr prediction requires some 200 time steps and approximately 10,000,000 multiplications for an area the size of the Earth's surface. The computing time on a machine with a multiplication speed of 100 μs, an addition speed of 10 μs, and a memory access time of 10 μs would be about 30 min. The magnitude of the computational task may be comprehended from the fact that the more accurate atmospheric models now envisaged will require some 100—1000 times this amount of computation.

A saving of time is accomplished by utilizing the quasi-non-divergent property of the large-scale atmospheric motions. If, in the above example, the horizontal divergence $\frac{\partial u}{\partial x} + \frac{\partial v}{\partial y}$ is set equal to zero, the motion is found to be completely described by the equation for the conservation of the vertical component of absolute vorticity.

The solution of this equation may be obtained in far fewer time steps since gravity wave motions are filtered out by this constraint and the velocity c in the Couranr-Friedrichs-Lewy criterion becomes merely the maximum particle velocity instead of the much greater gravity wave speed.

Cloud and Precipitation Prediction

If, to the standard dynamic variables u, v, ω, p, and ρ, a sixth variable, the density of water vapor, is added, it becomes possible to predict clouds and precipitation as well as the air motion. When a parcel of air containing a fixed quantity of water vapor ascends, it expands adiabatically and cools until it becomes saturated.

Continued ascent produces clouds and precipitation.

To incorporate these effects into a numerical prediction scheme one adds Eq. (5), which governs the rate of change of specific humidity r.

$$\frac{Dr}{Dt} \equiv \frac{\partial r}{\partial t} + u\frac{\partial r}{\partial x} + v\frac{\partial r}{\partial y} + \omega\frac{\partial r}{\partial z} = S \tag{5}$$

Here S represents a source or sink of moisture. Then it is necessary also to include as a heat source in the thermo-dynamic energy equation a term which represents the time rate of release of the latent heat of condensation of water vapor. The most successful predictions made by this method are obtained in regions of strong rising motion, whether induced by forced orographic ascent or by horizontal convergence in well-developed depressions. The physics and mechanics of the convective cloud-formation process make the prediction of convective cloud and showery precipitation more difficult.

Extended-Range Numerical Prediction

The extension of numerical predictions to long time intervals requires a more accurate knowledge than now exists of the energy transfer and turbulent dissipative processes within the atmosphere and at the air-earth boundary, as well as greatly augmented computing-machine speeds and capacities. However, predictions of mean conditions over large areas may well become possible before such developments have taken place, for it is now possible to incorporate into the prediction equations estimates of the energy sources and sinks——estimates which may be inaccurate in detail but correct in the mean. Several mathematical experiments involving such simplified energy sources have yielded predictions of mean circulations that strongly resemble those of the atmosphere.

Numerical Calculation of Climate

The above-mentioned experiments lead to a hope that it will be possible to explain the principal features of the earth's climate, that is, the average state of the weather, well before it becomes possible to predict the daily fluctuations of weather for extended periods. Should these hopes be realized it would then become possible to undertake a rational analysis of paleoclimatic variation and changes induced by artificial means. If the existing climate could be understood from a knowledge of the existing energy sources, atmospheric constituents, and earth surface characteristics, it might also be possible to predict the effects on the climate of natural or artificial modifications in one or more of these elements.

New Words

differential	[difə'renʃəl]	a. 微分的,差别的
succession	[sək'seʃən]	n. 连续、相继
infinitesimal	[infini'tesiməl]	a. 无穷小的
integration	[inti'greiʃən]	n. 积分,结合
nonlinearity	[,nɔnlini'æriti]	n. 非线性
complexity	[kəm'pleksiti]	n. 复杂性
multiplicity	[mʌlti'plisiti]	n. 多重性,复杂性
resort	[ri'zɔ:t]	v. 求助,凭借
grid	[grid]	n. 网格,地图坐标方格
span	[spæn]	n. 一段时间,v. 跨越
dream	[dri:m]	n. 梦
arithmetic	[ə'riθmətik]	n.a. 算术(的)
feasible	['fi:zəbl]	a. 可行的
prognosis	[prɔg'nəusis]	n. 诊断,预测
ignore	[ig'nɔ:]	v. 忽视,不顾
quasi-static	['kweisai'stætik]	a. 准静力(平衡)的
nondivergent	[,nɔndai'və:dʒən:t]	a. 无辐散的
filter	['filtə]	v. 滤去
amenable	[ə'mi:nəbl]	a. 适宜于,服从于
rectangular	[rek'tæŋgjulə]	a. 直角的,矩形的
coordinate	[kəu'ɔ:dineit]	n. 坐标
subscript	['sʌbskript]	n. 下标
derivative	[di'rivətiv]	n. 微商、导数
quotient	['kwəuʃənt]	n. 商
iterative	['itərətiv]	a. 重复的
criterion	[krai'tiəriən]	n. 标准,判据,判别式
criteria(复)		
memory	[meməri]	n. 记忆
comprehend	[kɔmpri'hend]	v. 理解,领会
envisage	[in'vizidʒ]	v. 想象,展望
constraint	[kən'streint]	n. 强制,约束
parcel	['pa:sl]	n. 一包,一块
adiabatically	[ædiə'bætikəli]	ad. 绝热地
incorporate	[in'kɔ:pəreit]	v. 把…列入…
orographic	[ɔrəu'græfik]	a. 地形的,山形的

depression	[diˈpreʃn]	n. 低压，凹地
paleoclimatic	[ˈpæliəuˈklaimitik]	a. 古气候的
undertake	[ˌʌndəˈteik]	v. 进行，从事，承担
filter out	[ˈfiltə]	滤去
partial differential equations	[ˈpɑːʃəl][ˌdifəˈrenʃəl][iˈkweiʃən]	偏微分方程
finite-difference approximation	[ˈfainait-ˈdifərəns][əˌprɔksiˈmeiʃən]	有限差分逼近
forced orographic ascent	[fɔːst][ˌɔrəuˈgræfik][əˈsent]	地形强迫上升
memory access	[ˈmeməri][ˈækses]	记忆存取
central difference quotient	[senˈtrɑːl][ˈdifərəns][ˈkwəuʃənt]	中央差商

24. Weather Forecasting and Its Accuracy

In most instances weather prediction is the ultimate goal of atmospheric research, the exceptions being the rather rare attempts at controlling the weather. It is also the most advanced area in meteorology. Because of the complex and highly quantitative nature of modern weather forecasting, we can only highlight the approaches used here, which include but are not limited to the traditional synoptic approach, statistical methods, and numerical weather prediction. The object of each of these methods is to not only project the location and possible intensification of existing pressure systems but also to determine the formation of new storm centers.

Synoptic weather forecasting was the primary method used in making weather predictions until the late 1950s. As the name implies, synoptic weather charts are the basis of these forecasts. From the careful study of weather charts over many years a set of empirical rules was established to aid the forecaster in estimating the rate and direction of weather system movements. For example, when the forecaster knows the type of weather being generated along a front and is able to predict its motion, a rather accurate forecast for the affected area can be made, but since cyclonic systems change so quickly, these forecasts are generally accurate only on a short-range basis of a few hours, or perhaps a day.

Early attempts to predict cyclone development from synoptic charts relied heavily upon the analysis of surface fronts. However, since the discovery of the relationship between the flow aloft and surface weather, these efforts have been supplanted by the use of upper-air data. Recall, for example, that wave cyclones can develop without the prior existence of surface fronts. More recently it has been shown that other methods can more accurately predict the future state of the atmosphere than can be accomplished by synoptic analysis. This is particularly true for forecasts made for periods longer than 1 or 2 days. Nevertheless, empirical rules applied to synoptic charts are still used by local forecasters in their attempt to pinpoint the occurrence of specific events such as the arrival time of a storm.

Modern weather forecasting relies heavily upon numerical weather prediction (NWP). The word "numerical" is misleading since all types of weather forecasting are based on some quantitative data and therefore could fit under this heading. The basis for NWP rests on the fact that the gases of the atmosphere obey a number of known physical principles. Ideally, these physical laws can be used to predict the future state of the atmosphere given the current conditions. This is analogous to predicting future positions of the moon based upon physical laws and the knowledge of its current position. However, the large number of variables that must be accounted for when considering the dynamic atmosphere makes this task extremely difficult. In order to simplify the problem, numerical models were developed which omit some of the variables by assuming that certain aspects of the atmosphere do not change with time. Although these models do not fully represent the "real" atmosphere, their usefulness in prediction has been well established. Most of these modern approaches strive to predict the flow pattern aloft. From this information meteorologists project favorable sites for cyclogenesis. However, even the most simiplfied models require such a vast number of calculations to be performed that they could only be used after the advent of high-speed computers. As faster computers are developed and improvements in atmospheric models are made, we can expect even greater advances in the area of weather prediction. Only a decade ago it was not uncommon to hear people say, "It should be nice tomorrow; the weather report calls for rain." This attitude no longer prevails, thanks partly to advances in NWP.

Although the accuracy of NWP had been shown to greatly exceed results obtained from more traditional methods, the prognostic charts obtained by these techniques are rather general. Hence, the detailed aspects of the weather must still be determined by applying traditional methods to these charts. Furthermore, numerical forecasts are limited by deficiencies in observational data. Stated another way, an NWP can be no more complete nor more accurate than the data that go into making it.

Statistical methods are often used in conjunction with and to supplement the NWP. Statistical procedures involve the study of past weather data in order to uncover those aspects of the weather that are good predictors of future events. Once these relationships are established, current data can then be used to predict future conditions. This procedure can be used to predict the overall weather, but it is most often used to determine one aspect of the weather at a time. For example, it is frequently used to project the maximum temperature for the day at a given location. This is done by first compiling statistical data relating temperature to wind speed and direction, cloud cover, humidity, and to the season of the year. These data are displayed on charts that provide a reasonable estimate of the maximum temperature for the day from these aspects of the current conditions.

Long-range forecasting is another area in which statistical studies have proven valuable. Currently, the National Weather Service prepares weekly and monthly weather outlooks. These are not weather forecasts in the usual sense; they are estimates of the rainfall and temperatures that should be expected during these periods. These projections only indicate whether or not the region will experience near normal conditions. Detailed forecasts for more than a few days are currently beyond the capability of the National Weather Service.

The general monthly extended forecasts are produced by first constructing a mean 700-hPa contour chart for the coming month. This requires taking into account the statistical records for that season of the year and altering them based upon the known effects of such things as ocean temperatures and snow cover. Once this chart is compiled, the relationships between the flow aloft and the development and movement of surface weather patterns are considered in making a prediction for each segment of the United States.

Another statistical approach to weather forecasting is called the analog method. The idea here is to locate in the weather records conditions that are as nearly analogous to the current conditions as possible. Once these are found, the sequence of weather events should parallel those of the past situation. Although this seems to be a very straightforward method of prediction, it is not without its drawbacks. No two periods of weather are identical in all respects, and there are just too many variables to match. Even when two periods seem to match very well, the sequence of weather that follows may be very different in each case. The main problem with this method may well be the lack of complete enough information, which as you recall greatly limited the usefulness of numerical weather predictions.

In summary, weather forecasts are produced by using several methods. The National Weather Service primarily uses NWP methods to generate large-scale prognostic charts. These charts are then disseminated to regional and local forecast centers that use traditional synoptic and statistical methods to generate more specific forecasts.

Forecast Accuracy

At one time or another most people have asked the question, "Why are the weather forecasts that are given on radio and television so often in error?" As contradictory as it might sound, the answer lies to some degree in our desire for more accurate predictions. When weather forecasting was in its infancy, the public was pleased when even an occasional forecast came true. Today, we expect more accurate predictions, and hence we concentrate more on the relatively few incorrect forecasts and take for granted the more accurate predictions. Also, as weather forecasting became more accurate, attempts were made to make the forecasts even more specific. Instead of just predicting the likelihood of precipitation, a modern forecast often gives the expected times of occurrence and the probable amount. Just the other day I overheard a colleague complain that the forecast called for 3 to 6 inches of snow, and we got only an inch. At one time this forecast would have been considered correct, but today it is viewed, at least by some, as inaccurate.

In general, we are safe in saying that modern weather forecasts are relatively accurate for the immediate future (from 6 to 12 hours), and they are generally good for a day or two. However, after a forecast becomes a few days old, its accuracy decreases rapidly. Even today, specific forecasts are only made for periods of 3 to 5 days in advance and they are constantly being revised.

When making very short-range predictions of a few hours, it is difficult to improve on persistence forecasts. These forecasts assume that the weather occurring upstream will persist and move on and will affect the areas in its path in much the same way. Persistence forecasts are used by local forecasters in determining such events as the time of the arrival of a thunderstorm that is moving toward their region. Persistence forecasts do not account for changes that might occur in the intensity or in the path of a weather system, and they do not predict the formation or dissipation of cyclones. Because of these limitations, and the rapidity with which weather systems change, persistence forecasts break down after 12 hours, or a day at most.

For periods of up to 5 days forecasts made by numerical weather prediction methods and supplemented by traditional techniques are difficult to beat. They take into consideration both the formation and the movement of pressure systems. However, beyond 5 days specific forecasts made by the more sophisticated methods prove to be no more accurate than projections made from past climatic data. The reasons for the limited range of modern forecasting techniques are many. As stated earlier, the network of observing stations is rather incomplete. Not only are large areas of the earth's land-sea surface inadequately monitored, but data gathering in the middle and upper troposphere is meager at best, except in a few regions. Further, the laws governing the atmosphere are not completely understood and the current models of the atmosphere are not as complete as possible. Nevertheless, NWP has greatly improved the forecaster's ability to project changes in the upper-level flow. When the flow aloft can be tied more fully to surface conditions, weather forecasting should improve greatly.

You may have asked, "Just how accurately is the weather currently being forecast?" This question is more difficult to answer than you might imagine. One of the major problems is establishing when forecast is correct. For example, if a forecast predicts a minimum temperature of 10℃ and the temperature fails to 9℃, only 1 degree off, is that forecast incorrect? When a forecast calls for snow in Wisconsin and only the northern two-thirds of the state receives snow, is that forecast incorrect or is it two-thirds correct? The problems of assessing forecast accuracy are many, as you can see from these examples.

The only aspect of the weather that is predicted as a percent probability is rainfall. Here statistical data are used to indicate the number of times precipitation occurred under similar conditions. Although the occurrence of precipitation can be predicted with about 80 percent accuracy, predictions on the amount and time of occurrence of precipitation are still fairly unreliable. Probably temperatures and wind directions can be most accurately predicted.

New Words

instance	['instəns]	n.	情况,实例
goal	[gəul]	n.	目标,目的
exception	[ik'sepʃən]	n.	例外
traditional	[træ'diʃənl]	a.	传统的
statistical	[stə'tistikəl]	a.	统计(学)的
intensification	[intensifi'keiʃn]	n.	强化;增强
supplant	[sə'plɑːnt]	v.	取代
recall	[ri'kɔːl]	v.	回想,想起
prior	['praiə]	a.	先前的,居先的
pinpoint	['pinpɔint]	v.	正确指出,准确定位
mislead	[mis'liːd]	v.	使误解
analogous	[ə'næləgəs]	a.	类似的,模拟的
omit	[əu'mit]	v.	省略,遗漏
strive	[straiv]	v.	努力
attitude	['ætitjuːd]	n.	态度,姿态
prognostic	[prɔg'nɔstik]	a.	预报的,预兆的
deficiency	[di'fiʃənsi]	n.	缺乏,不足

conjunction	[kən'dʒʌŋkʃən]	n. 连接, 结合
supplement	['sʌplimənt]	v. n. 补充
compile	[kəm'pail]	v. 编辑, 整编
outlook	['autluk]	n. 展望, 前景
contour	['kɔntuə]	n. 等值线
segment	['segmənt]	n. 部分; 段; 节
analog	['ænələg]	n. 类似(物), 模拟量
sequence	['si:kwəns]	n. 顺序, 序列
straightforward	[streit'fɔ:wəd]	a. 简单的, 直接的
drawback	['drɔ:bæk]	n. 弊端, 欠缺
summary	['sʌməri]	n. 提要, 摘要
disseminate	[di'semineit]	v. 传播, 散布
error	['erə]	n. 错误, 误差
contradictory	[kɔntrə'diktəri]	a. 矛盾的, 对立的
infancy	['infənsi]	a. 初期, 幼年时期
likelihood	[laiklihud]	n. 可能性; 相似性
overhear	[əuvə'hiə]	v. 偶然听到, 偷听
colleague	['kɔli:g]	n. 同事
complain	[kəm'plein]	v. 抱怨
revise	[ri'vaiz]	v. 修正, 修订
persistence	[pə'sistəns]	n. 持续
upstream	['ʌp'stri:m]	ad. 向上游
limitation	[limi'teiʃn]	n. 局限性, 限制
sophisticated	[sə'fistikeitid]	a. 成熟的, 完善的
imagine	[i'mædʒin]	v. 想象, 推测
assess	[ə'ses]	v. 评价, 确定
probability	[prɔbə'biliti]	n. 概率, 可能性
in conjunction with		连同
take … into account		考虑
in summary		概要说来
take … for granted		认为…是理所当然的
break down		破环, 发生故障, 分解

Supplemental reading 24.1

Atmospheric motions are inherently unpredictable as an initial value problem (i.e as a system of equations integrated forward in time from specified initial conditions) beyond a few weeks. Beyond that time frame, uncertainties in the forecasts, no matter how small might be in the initial conditions, because as large as the observed variations in atmospheric flow patterns. Such exquisite sensitivity to initial conditions is characteristic of a broad class of mathematical models of real phenomena, referred to as chaotic nonlinear systems. In fact, it was the growth of errors in a highly simplified weather forecast model that provided one of the most lucid early demonstrations of this type of behavior.

In 1960, Professor Edward N Lorenz in the Department of Meteorology at MIT decided to rerun an experiment with a simplified atmospheric model in order to extend his "weather forecast" father out into the future. To his surprise, he found that he was unable to duplicate his previous forecast. Even though the code and the prescribed initial conditions in the two experiments were identical, the states of the model in the two forecasts diverged, over the course of the first few hundred time steps, to the point that they were no more like one another than randomly chosen states in experiments started from entirely different initial conditions. Lorenz eventually discovered the computer he was using was introducing round-off errors in the last significant digit that were different each time he ran the experiment. Differences between the "weather patterns" in the different runs were virtually indistinguishable at first, but they grew with each time step until they eventually became as large as the range of variations in the individual model runs.

Lorenz's model exhibited another distinctive and quite unexpected form of behavior. For long periods of (simulated) time it would oscillate around some "climatological-mean" state. Then, for no apparent reason, the state of the model would undergo an abrupt "regime shift" and begin to oscillate around another quite different state, as illustrated in Fig. 1. Lorenz's model exhibited two such preferred "climate regimes". When the state of the model resided within one of these regimes, the "weather" exhibited quasi-periodic oscillations and consequently was predictable quite far into the future. However, the shifts between regimes were abrupt, irregular, and inherently unpredictable beyond a few simulated days. Lorenz referred to the two climates in the model as *attractors*.

The behavior of the real atmosphere is much more complicated than that of the highly simplified model used by Lorenz in his experiments. Whether the Earth's climate exhibits such regime-like behavior, with multiple "attractors", or weather it should be viewed as varying about a single state that varies in time in response to solar, orbital, volcanic, and anthropogenic forcing is a matter of ongoing debate. (Adapted from Wallace et al., 2006).

Fig. 1 The history of the state of the model used by Lorenz can be represented as a trajectory in a three-dimensional space defined by the amplitudes of the model's three dependent variables. Regime-like behavior is clearly apparent in this rendition. Oscillations around the two different "climate attractors" correspond to the two, distinctly different sets of spirals, which lie in two different planes in the three dimensional phase space. Transitions between the two regimes occur relatively infrequently

Supplemental reading 24.2

The severe thunderstorm forecast and warning lead-time, defined from its issuance to the first severe weather report, has quite a bit of variability between the respondent countries. Of the 28 countries that forecast severe thunder-storms, 8 (29%) may issue their forecasts more than 24 h before the event.

25. Long-Range Weather Prediction

(1) Definitions of Scope

(A) Time

"Long-range" weather forecasting begins after the sequence of ordinary day-by-day forecasts has lost its margin of accuracy above simple persistence. This may happen at any range from 3 to 6 days, depending on the place, the season, and the current form of the general circulation of the atmosphere. Beyond the variable cutoff, predictions must be confined to the more general statistical properties of the weather, e. g., mean temperature or total precipitation. The shortest period of averaging that will smooth out the daily fluctuations is about five days. Predictions of average over such periods, particularly if they are timed to begin soon after being made and thus to include some short-range information, are more properly known as "extended" forecasts. They are coming to depend (in the U. S. at least) more and more on the new methods of short-range forecasting by direct numerical solution of the equations governing atmospheric motions. These methods, and the electronic computers required to use them have not yet been developed to the stage where they might produce forecasts of the departures from normal of the average weather for a couple of weeks, a month, or a season——that is, true long-range forecasts. Whether or not they can be remains at present an open question of great scientific and practical interest, to which we shall return in Section (3). Variations of annual or even longer-term averages belong to the realm of "climatic fluctuations". Forecasts of them are seldom attempted.

(B) Space

Long-range forecasts may be aimed at specific locations (such as cities), geographical regions, nations, or even the whole northern hemisphere, depending on the user's requirements. A single area-average is sometimes given for regions not exceeding a few thousand square miles, but a forecast map showing spatial variations is needed for larger areas.

A few of the forecasting methods now in use draw only on local measurements for their input data, but most employ data from a much larger area than the predictions will cover, for it is known that local weather can be affected within a few days by meteorological conditions thousand of miles away.

(2) Current Practice

(A) Countries

According to a recent survey, extended or long-range forecasts are being made by the national weather services of the United States, Great Britain, France, West Germany, Italy, Sweden, and Turkey; the Soviet Union, East Germany, and Hungary; and India, Pakistan, Thailand, Indonesia, and Japan. It is probable that the list is incomplete. In addition, private individuals and groups in some of these countries produce forecasts independently. The dissemination of forecasts varies greatly in mode and breadth from one nation to the next; some are published in full, others are restricted to a few official uses.

(B) Methods

Because the long-range forecast problem has obdurately resisted all attempts at solution by quantitative physical theory, current methods of prediction——the modest survivors from a long history of generally fruitless research——are largely empirical in nature. The principal methods are:

(a) Statistical regression analysis and related techniques, based on time-lagged correlations between one or more predictor variables and the predictand, or on the autocorrelations of the predictand.

(b) Contingency tables, in which the whole possible range of the predictand is divided into a few broad classes, the predictors also classified, and the frequency of occurrence of each predictand class contingent on the prior occurrence of each possible combination of predictor classes found.

(c) Forward extrapolation of apparent periodicities or trends in the predictand record.

(d) Kinematic extrapolation of recent rates of movement or of intensification of features on maps of the predictand.

(e) Selection from the past of one or more analogs to the present situation, i. e., closely similar cases. The prediction may follow either the average or the most frequent subsequent development of the analog cases.

(f) Forecaster's judgement, based on experience.

(g) Forecaster's judgement, based on qualitative physical reasoning from current theoretical concepts.

Each of the above methods often has been abused, and each requires very cautious application to produce useful results. Only methods(a), (b) and (d) are wholly objective, although (c) and (e) can be made so. Most of the forecast services use a combination of several.

(C) Accuracy

Predictions are said to have positive "skill" if, on the average, they show greater accuracy than would a series of control forecasts generated by random drawing from the climatological frequency distribution of the predictand. Forecasts are not generally considered to be useful unless they also maintain some margin of accuracy above another set of control forecasts generated by simple persistence (repetition) of the preceding value of the predictand.

Because of the great variety of the ways in which long-range forecasts are expressed and of the uses to which they are put, it is difficult to rate their skill adequately or fairly with single numbers. If one must attempt it nevertheless, the simplest common denominator is perhaps a score based on an even two-way choice as, for instance, above or blow normal. If all long-range forecasts were to be reduced to these terms, one would find the U. S. five-days forecasts of temperature to be right about 70% of the time on the average, and as much as 85% of the time in favorable circumstances. For monthly temperature forecasts in the U. S., Great Britain, and West Germany, the comparable figures would be about 60% on the average and 65% at best. Experimental (unpublished) seasonal temperature forecasts in the U. S. would score about 55% on the average and 60% under some conditions. All of these scores are higher than could be obtained by predicting temperatures of the previous period to persist unchanged. It does not, however, appear to be possible at present to forecast total precipitation for a month or season with an average score of more than 53%. The corresponding figures for five days in the U. S. would be 56% on the average and

60% at best. A contributing reason for these lower scores is the fact that precipitation occurs in more complex and broken spatial patterns than temperature, i. e., local precipitation totals over a period are sensitive to the paths and intensities of individual storms to a far greater degree than are average temperatures.

(3) Physical Theory

Some physical concepts, such as the length of stable waves in the westerly winds aloft which circle the globe in middle latitudes, the cooling effects of anomalous snow cover, and the contribution of unusually warm sea-surface temperatures to increased cyclogenesis, are employed to a degree in long-range forecasting qualitatively. An accepted quantitative physical theory of long-range forecasting cannot, however, be said to exist.

There does exist a family of mathematical models of the atmosphere, incorporating (with various simplifications and omissions) the differential equations for the conservation of momentum, energy and mass, plus the equation of state. Given initial conditions and boundary conditions, one can integrate a set of these equations numerically with respect to time, obtaining a step-by-step (or iterative) sequence of predictions. This kind of direct application of classical fluid dynamics and thermodynamics is being successfully used in some aspects of short-range and extended forecasting. With the addition of radiative and frictional processes, and after many time iterations, one gets a dynamical prediction of certain aspects of climate such as the normal wind and temperature fields, normal transport and transformations of momentum and energy. This line of research is currently undergoing rapid elaboration and refinement.

The models of the general circulation, or some further development from them, would seem to provide natural generators of long-range forecasts. They are still incomplete in a physical sense, coarse in spatial resolution, and subject to some cumulative computational errors, but these are deficiencies which experience and faster computers will gradually overcome. More difficult problems arise with respect to the boundary and initial conditions. Some of the boundary conditions cannot be specified in advance, but may depend in part on the instantaneous or recent internal state of the atmosphere; good examples are the dependence of the radiation balance on cloudiness and on snow cover. Such variable feedback effects require very careful modeling if the computed system is to remain stable and produce realistic results. But even the most sophisticated model must begin with initial conditions, which cannot be but measured everywhere or with absolute accuracy. This inevitable incompleteness of knowledge of the state of the atmosphere at any one moment will probably cause a dynamical prediction beginning at that moment of lose all of its accuracy with respect to daily evolution after a week or two. The practical consequence of this result is the empirically long-recognized requirement that long-range forecasts deal only with gross statistics of the weather, such as average values and probability distributions. The crucial question then becomes: Will the evolutions of two nearly identical initial states of the atmosphere diverge in gross statistics as well as in detail, and if so, how rapidly? Forecasters have naturally assumed a slow rate of divergence; this amounts to believing that a "good analog", if found, will lead to a reasonably good prediction of mean conditions. Equivalently, it implies that practical long-range forecasting of these mean conditions by dynamical models is possible. By studying the behavior of a highly simplified nonlinear model, however, Lorenz (1963) has found that even in such a system, gross statistical divergence of two nearly identical initial states will occur, sometimes in a week. The corresponding time scale for the real atmosphere is not known. This result seems to carry grave implications of long-range forecasting. On the other hand, the fact that various kinds of empiricism have provided forecasts of demonstrated (though modest) skill means either that the time scale of Lorenz's kind of instability is rather long or else that influences external to the atmosphere tend to force its evolution in some preferred direction. Such forcing functions do not remain constant: they are only required to vary slowly enough to remain fairly predictable through the period in question. One obvious example is the elevation angle of the sun, which produces the normal march of the seasons. Another, relevant to the forecasting problem of departures from the normal, is the field of sea-surface temperature anomalies. Such forcing functions can be incorporated, as they are identified, into the models of the general circulation as part of the boundary conditions and thus contribute a degree of predictability through their influence on the dynamics of the system. Alternatively, it may prove possible to construct other models, centered on these thermodynamic forcing functions and possessing simpler dynamical properties, which still retain valid predictive power. Recent studies by Adem, Namias and Saltzman lead in this direction.

We may conclude from the foregoing discussion, then, that it is probable that research in the coming decade will disclose the future of long-range forecasting theory and practice.

New Words

cutoff	[ˈkʌtɔf]	n.	切断、割去
confine	[kənˈfain]	v.	限制,局限
smooth	[smuːð]	v. a.	使光滑,消除;光滑的,平坦的
realm	[relm]	n.	领域,范围
private	[ˈpraivit]	a.	私人的
breadth	[braedθ]	n.	宽度
official	[əˈfiʃəl]	a.	官方的,正式的
obdurately	[ˈɔbdjuritli]	ad.	执拗地
modest	[ˈmɔdist]	a.	适度的
survivor	[səˈvaivə]	n.	幸存者
empirical	[emˈpirikl]	a.	经验的,实验的
regression	[riˈgreʃn]	n.	回归
predictor	[priˈdiktə]	n.	预报因子
predictand	[priˈdiktənd]	n.	预报量
autocorrelation	[ˈɔːtəukɔriˈleiʃn]	n.	自相关
contingency	[kənˈtindʒənsi]	n.	列联,偶发性
contingent	[kənˈtindʒənt]	a.	伴随的;视…而定
extrapolation	[ekstrəpəˈleiʃn]	n.	外推法
kinematic	[kainiˈmætik]	a.	运动学的
subsequent	[ˈsʌbsikwənt]	a.	后继的
abuse	[əˈbjuːs]	v.	滥用,妄用
repetition	[repiˈtiʃn]	n.	重复
skill	[skil]	n.	技巧
denominator	[diˈnɔmineitə]	n.	分母,标准
anomalous	[əˈnɔmɔləs]	a.	反常的,不规则的
iteration	[itəˈreiʃn]	n.	重复
coarse	[kɔːs]	a.	粗糙的
resolution	[rezəˈluːʃn]	n.	解决;决议;分辨率
instantaneous	[instənˈteinjəs]	a.	瞬间
feedback	[ˈfiːdbæk]	n.	反馈,回输
realistic	[riəˈlistik]	a.	现实的
gross	[grəus]	n. a.	总体,总的
crucial	[kruːʃəl]	a.	决定性的
nonlinear	[ˈnɔnˈliniə]	a.	非线性的
implication	[impliˈkeiʃn]	n.	含义,推断
empiricism	[emˈpirisizm]	n.	经验主义
relevant	[ˈrelivənt]	a.	与…有关的
alternatively	[ɔːlˈtəːnətivli]	ad.	选择地,另一方面
time-lagged	[taim-ˈlægid]		时间滞后的
control forecast	[kənˈtrəul][ˈfɔːkaːst]		对比预报

in advance [əd'vɑːns] 预先,提前

Proper Nouns

Britain	['britin]	英国
France	[frɑːns]	法国
Germany	['dʒəːməni]	德国
Italy	['itəli]	意大利
Sweden	['swiːdn]	瑞典
Turkey	['təːki]	土耳其
Hungary	[hʌŋɡər]	匈牙利
India	['indjə]	印度
Pakistan	[pɑːkisˈtæn]	巴基斯坦
Tailand	['tailænd]	泰国
Indonesia	[indəuˈniːzjə]	印度尼西亚
Soviet Union	['səuvietˈjuːnjən]	苏联

26. Introducing the New Generation of Chinese Geostationary Weather Satellites, Fengyun-4

Fengyun-4 is the new generation of Chinese geostationary meteorological satellites with greatly enhanced capabilities for high-impact weather event monitoring, warning, and forecasting.

On 7 September 1988, the Long March rocket carried the Fengyun-1A (FY-1A) polar-orbiting satellite into orbit, marking the start of the Chinese Fengyun (FY; meaning wind and cloud in Chinese) meteorological satellite observing systems program. On 10 June 1997, the FY-2A geostationary satellite was successfully launched and the Chinese meteorological satellite program took a large step toward its goal of establishing both polar-orbiting and geostationary observational systems. Over time the Fengyun satellites have become increasingly important for protecting lives and property from natural disasters in China.

The Chinese FY satellites are launched as a series. The odd numbers denote the polar-orbiting satellite series, and the even numbers denote the geostationary satellite series. After launch, a letter is appended to indicate the order in the satellite series; for instance, FY-2F is the sixth satellite that has been launched in the first generation of the Chinese geostationary satellites (FY-2). The FY-2, the first generation of the Fengyun geostationary weather satellite series, includes seven satellites launched since 1997; another will follow around 2017 to conclude the FY-2 mission. The Chinese geostationary weather satellite system operates two satellites located at 86.5°E (FY-2 West) and 105°E (FY-2 East); they provide full-disc observations every 30 min and observations every 15 min in their overlap region (see Fig. 1 for the current FY-2 coverage). The FY-2D (West) and FY-2E (East) observation schedule is shown in Table 1. The two satellites also back each other up. FY-2 satellites carry the Visible and Infrared Spin Scan Radiometer (VISSR) capable of imagery in five spectral bands. Derived products include atmospheric motion vectors (AMVs), sea surface temperatures (SSTs), total precipitable water vapor (TPW), quantitative precipitation estimations (QPEs), fire locations and intensity, surface albedo, and several others.

Table 1　FY-2D/FY-2E overlap region observation schedule. Full-disc and north-disc observations are abbreviated as F and N, respectively

Time (UTC)	Region	Satellite
0000	F	FY-2E
0015	F	FY-2D
0030	N	FY-2E
0045	N	FY-2D
0100	F	FY-2E
0115	F	FY-2D
0130	N	FY-2E
0145	N	FY-2D
—	—	—
2300	F	FY-2E
2315	F	FY-2D
2330	N	FY-2E
2345	N	FY-2D

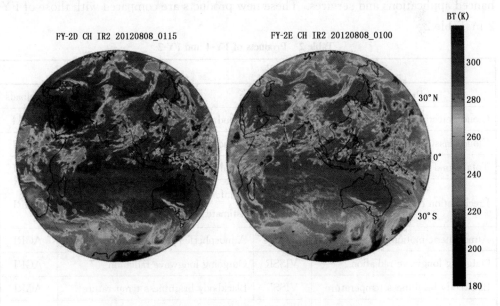

Fig. 1　The 10.8-μm (left) FY-2D and (right) FY-2E BT (K) images showing coverage of (left) FY-2 East and (right) FY-2 West

The FY-4 introduces a new generation of Chinese geostationary meteorological satellites, with the first FY-4A launched on 11 December 2016. The remaining satellites of this series are planned to be launched from 2018 to 2025 and beyond. FY-4 has improved capabilities for weather and environmental monitoring, including a new capability for vertical temperature and moisture sounding of the atmosphere with its high-spectral-resolution infrared (IR) sounder, the Geostationary Interferometric Infrared Sounder (GIIRS). Following 15 years, the three-axis stabilized FY-4 series will offer full-disc coverage every 15 min or better (compared to 30 min of FY-2) and the option for more rapid regional and mesoscale observation modes. The Advanced Geosynchronous Radiation Imager (AGRI) has 14 spectral bands (increased from the five bands of FY-2) that are quantized with 12 bits per pixel (up from 10 bits for FY-2) and sampled at 1 km at nadir in the visible (VIS), 2 km in the near-infrared (NIR), and 4 km in the remaining IR spectral bands (compared with 1.25 km for VIS, no NIR, and 5 km for IR of FY-2). FY-4 will improve most products of FY-2 and introduce many new products [such as atmospheric temperature and moisture profiles, atmospheric instability indices, layer precipitable water vapor (LPW), rapid developing clouds, and others]. Products from FY4 series are expected to provide enhanced applications and services. These new products are compared with those of FY-2 in Table 2.

Table 2 Products of FY-4 and FY-2

FY-2		FY-4	
Products	Payloads	Products	Payloads
Cloud detection	VISSR	Cloud masks	AGRI
Cloud classification	VISSR	Cloud type	AGRI
Total cloud amount	VISSR	Total cloud amount	AGRI
Precipitation estimation	VISSR	Rainfall rate/quantitative precipitation estimate	AGRI
Atmospheric motion vector	VISSR	Atmospheric motion vector	AGRI
Outgoing longwave radiation	VISSR	Outgoing longwave radiation	AGRI
Blackbody brightness temperature	VISSR	Blackbody brightness temperature	AGRI
Surface solar irradiance	VISSR	Surface solar irradiance	AGRI

continued

FY-2		FY-4	
Products	Payloads	Products	Payloads
Humidity product analyzed by cloud information	VISSR	Legacy vertical moisture profile	GIIRS
Total precipitable water	VISSR	Layer precipitable water	AGRI
Upper-tropospheric humidity	VISSR	Layer precipitable water	
Dust detection	VISSR	Aerosol detection (including smoke and dust)	AGRI
Sea surface temperature	VISSR	Sea surface temperature (skin)	AGRI
Snow cover	VISSR	Snow cover	AGRI
Land surface temperature	VISSR	Land surface (skin) temperature	AGRI
Cloud-top temperature	VISSR	Cloud-top temperature	AGRI
		Cloud-top height	AGRI
		Cloud-top pressure	AGRI
		Cloud optical depth	AGRI
		Cloud liquid water	AGRI
		Cloud particle size distribution	AGRI
		Cloud phase	AGRI
		Downward longwave radiation: surface	AGRI
		Upward longwave radiation: surface	AGRI
		Reflected shortwave radiation: top of atmosphere	AGRI
		Aerosol optical depth	AGRI
		Convective initiation	AGRI
		Fire/hot spot characterization	AGRI
		Fog detection	AGRI
		Land surface emissivity	AGRI
		Land surface temperature	AGRI

	FY-2	FY-4	continued
Products	Payloads	Products	Payloads
		Land surface albedo	AGRI
		Tropopause folding turbulence prediction	AGRI
		Legacy vertical temperature profile	GIIRS
		Ozone profile and total	GIIRS
		Atmosphere instability index	GIIRS
		Lightning detection	LMI
		Space and solar products	SEP

FY-4's AGRI will be operated in conjunction with GIIRS. FY-4's GIIRS is the first high-spectral-resolution advanced IR sounder on board a geostationary weather satellite, complementing the advanced IR sounders in polar orbit. These include the Atmospheric Infrared Sounder (AIRS) on board the National Aeronautics and Space Administration (NASA) Earth Observing System (EOS) Aqua platform (Chahine et al., 2006), the Infrared Atmospheric Sounding Interferometer (IASI) on board Europe's Meteorological Operational (MetOp) satellites (Clerbaux et al., 2007; Smith et al., 2009), and the Cross-Track Infrared Sounder (CrIS) on board the Suomi National Polar-Orbiting Partnership (SNPP; www.nasa.gov/mission_pages/NPP/main/index.html; Bloom 2001). They have had a large positive impact in global and regional numerical weather prediction (NWP) applications (Le Marshall et al., 2006; McNally et al., 2014; Wang et al., 2014; Li et al., 2016) and climate research (Yoo et al., 2013). However, severe weather warning in a preconvective environment (Li et al., 2011, 2012), nowcasting, and short-range forecasting require nearly continuous monitoring of the vertical temperature and moisture structure of the atmosphere on small spatial scales that only a geostationary advanced IR sounder can provide. The GIIRS will provide breakthrough measurements with the temporal, horizontal, and vertical resolution needed to resolve the quickly changing water vapor and temperature structures associated with severe weather events. GIIRS will be an unprecedented source of information on the dynamic and thermodynamic atmospheric fields necessary for improved nowcasting and NWP services (Schmit et al., 2009). High-spectral-resolution IR measurements will also provide estimates of diurnal variations in tropospheric trace gases like ozone and carbon monoxide (Li et al., 2001; J.-L. Li et al., 2007; Huang et al., 2013) that will support forecasting of air quality and monitoring of atmospheric minor constituents.

The FY-4 GIIRS is one of the Group on Earth Observations (GEO) sounders planned by Global Earth Observation System of Systems (GEOSS) member states in response to the call from the World Meteorological Organization (WMO) for advanced sounders in the geostationary orbit. Another is the Infrared Sounder (IRS) planned by the European Organisation for the Exploitation of Meteorological Satellites (EUMETSAT) for the geosynchronous Meteosat Third Generation (MTG) satellite systems in the 2020 time frame and beyond. Together with the new generation of geostationary weather satellite systems being developed by other countries, FY-4 will become an important GEO component of the global Earth-observing system.

Overall, FY-4 represents an exciting expansion in Chinese geostationary remote sensing capabilities. FY-4A will be considered experimental and the subsequent satellites in the FY-4 series will be operational. Compared with the current operational FY-2 series, the FY-4 satellites are designed to have a longer operating life. Table 3 summarizes some of the significant improvements in instrument performance expected from FY-4 compared with the current operational FY-2 series. For the FY-4 operational series of satellites, the main observation capabilities are similar to those of FY-4A, with some significant performance improvements. The AGRI channel number will be increased from 14 to 18 with IR spatial resolution of 2 km, and the full-disc temporal resolution will be enhanced from 15 to 5 min. GIIRS spectral and spatial resolutions will be increased to 0.625 cm - 1 and 8 km, respectively. Lightning Mapping Imager (LMI) coverage will be enlarged to full disc. Space-monitoring instruments will be increased; for example, a solar X-ray and extreme ultraviolet imager will be on the following FY-4 series satellites.

Table 3 Advancement of FY-4A compared with the current operational FY-2 series. SEM = Space Environment Monitor. SSP = subsatellite point

	FY-4A (experimental)	FY-4 (operational)	FY-2 (operational)
Stabilization	Three axis	Three axis	Spin
Designed life	7 years (designed life)	7 years (operation life)	4 years
Observation efficiency	85%	85%	5%

continued

Observation mode	Imaging + sounding + lightning mapping	Imaging + sounding + lightning mapping	Imaging only
Main instruments	AGRI: 14 channels Resolution: 0.5—4 km Full disc: 15 min	AGRI: 18 channels Resolution: 0.5—2 km Full disc: 5 min	VISSR: 5 channels Resolution: 1.25—5 km Full disc: 30 min
	GIIRS: 913 channels SSP resolution: 16 km Spectral resolution: 0.8, 1.6 cm^{-1}	GIIRS: >1,500 channels SSP resolution: 8 km Spectral resolution: 0.625 cm^{-1}	—
	LMI Area coverage SSP resolution: 7.8 km	LMI Full-disc coverage SSP resolution: 7.8 km	—
	SEP High-energy particles Magnetic field	SEP High-energy particles, magnetic field, solar imager	SEM High-energy particles Solar X-ray fluxes

This paper provides an introduction to the Chinese FY-4 observation capabilities, the derived products, and the associated applications. The ground system components are briefly described in the next section. The following sections provide an overview of the four FY-4A instruments, products, and related application areas, and the final section offers a summary and conclusions.

FY-4A GROUND SEGMENT. A new ground segment has been designed and is being built to accommodate the technical requirements of the FY-4 satellites. The primary ground missions are as follows:

1) receiving raw data from the satellite;

2) determining and predicting the satellite orbit based on ranging measurements to the satellite;

3) monitoring the satellite and controlling the payloads;

4) undertaking the mission management and operation control of the satellite and ground systems;

5) processing data for geolocation and registration;

6) processing data for measurement calibration;

7) producing quantitative products;
8) providing an archive and distribution service for the data and products;
9) carrying out applications for the weather, climate, and environment; and
10) accomplishing monitoring and predicting services for space weather

Fig. 2 presents a flowchart of FY-4 ground segment. In addition to the backup Data and Telemetry System (DTS) located in Guangzhou in southern China, the new DTS facilities will be located in Beijing, China. The ground segment is being developed in five phases, which include user requirement analysis, algorithm development for quantitative products, system design, engineering, and in-orbit testing.

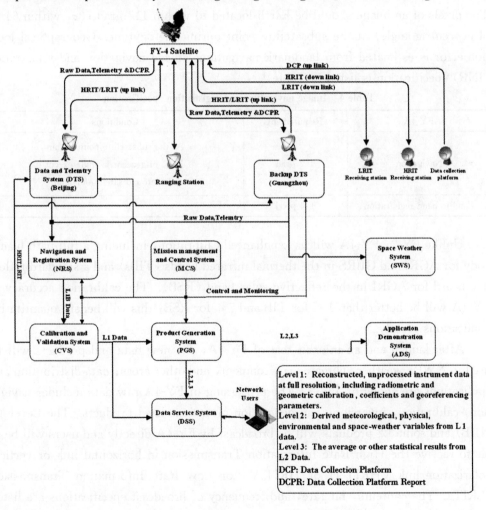

Fig. 2 Flowchart of FY-4 ground segment

The navigation and registration of AGRI, GIIRS, and LMI data from a three-axis stabilized satellite is a great challenge. A method has been designed in which the satellite platform, payloads, and ground segment cooperate with each other. As part of the ground segment, the Navigation and Registration System (NRS) calculates the scan and step angles from each instrument to the predicted stars that can be observed by the two payloads and arranges the observation timetable for AGRI and GIIRS based on those predicted stars. The NRS solves the equations that describe the relationship between the optical line of sight of instrument and structural thermal distortion and determines the equivalent variation of yaw, pitch, and roll once every 24 h. The pixels of an image should be Earth-located to within 112 μrad (3σ, within 64.5° of geocentric angle) at the subsatellite point during the daytime. Geographical location error is estimated from landmark navigation. Image navigation and registration (INR) specification is listed in Table 4.

Table 4 Image navigation and registration specification

FY-4A	Requirement	Conditions
Navigation	112 μrad	At the subsatellite point within 64.5° of geocentric angle except ±2 h around satellite midnight
Band-to-band registration	1/4 pixel	—

Unlike FY-2, FY-4A will have enhanced calibration, including a full-path blackbody for AGRI and GIIRS in the thermal infrared bands (TIBs) and a standard reflective board for AGRI in the reflective solar band (RSB). The calibration accuracy of FY-4A will be better than 1 K for TIB and 5% for RSB; this will benefit quantitative applications greatly.

After launch and an in-orbit test of FY-4A, the new data and products will be used in NWP, weather, climate, environment, and other areas; data distribution and applications are shown in Fig. 3. The processing of FY-4A raw data includes navigation, calibration, inversion, and generation of various-level products. The Level 1B (L1B) and some L2 products will be broadcast by FY-4A directly and users will be able to receive the High Rate Information Transmission in horizontal link or vertical polarization link (HRIT-H or HRIT-V) or Low Rate Information Transmission (LRIT). The contents, bit rate, and frequency of broadcast specifications are listed in Table 5. The L2 and L3 products generated by the ground segment of FY-4A will

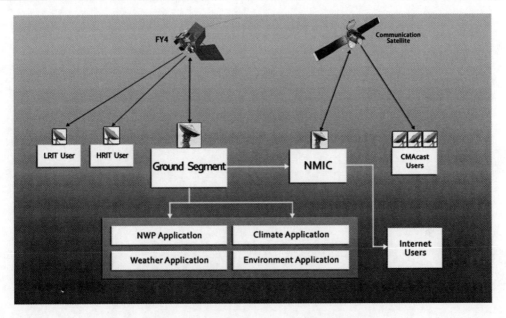

Fig. 3 FY-4A data distribution and application flowchart

also be distributed by the National Meteorological Information Center (NMIC) through CMACast (Satellite Data Broadcasting System of China Meteorological Administration). All datasets of FY-4A, both real time and historical, will be available to the global community on the National Satellite Meteorological Center (NSMC) satellite data server website (http://satellite.nsmc.org.cn).

Table 5 FY-4A direct broadcast capabilities

	Channel	Bit rate	Max daily data (GB)	Contents	Frequency
1	HRIT-H	11.6 Mbps	123	L1 of all 14 channel data of AGRI	1,680 MHz
2	HRIT-V1	9.3 Mbps	65.9	a) GIIRS data b) LMI data c) Part of L2 products	1,679 MHz
3	HRIT-V2	750 Kbps	2.6	A part of AGRI data	1,679 MHz
4	LRIT	150 Kbps	1.58	Low-resolution image of AGRI	1,697 MHz

(Editor's note: We are sorry for that we have to omit the remainder part of this excellent article since the length limitation of the book. Readers can find the whole article from American Meteorological Society, August 2017, BAMS:1637-1658(Yang et al.,2017).

第二部分 气象科技英语听说材料

Part Two: Listening and Spoken English Materials for Meteorological Science and Technology

1. A Weather Report

(1) Micro-listening

This is Radio Station WBRS in Seattle. At 9:00 it's time for the morning weather report, brought to you by McDonald's Hamburgers. This morning's skies will be cloudy, but no rain today. By noon the sun will be shining. The temperature is now 62 with a high of about 85 for this afternoon. A good day for swimming! Tomorrow's weather is next.

This is WBRS in Seattle with tomorrow's weather. Morning clouds with the sun in the afternoon. The low temperature tonight will be 60, with a high tomorrow of 86. Have a nice day! This weather report has been brought to you by McDonald's.

(2) Answer the following questions:

1) Is this weather report broadcast in the morning or in the afternoon?
2) Where is this radio station located?
3) It is not cloudy this morning, is it ?
4) What is the high temperature for tomorrow?
5) Is tomorrow a good day for swimming? Why?
6) Why is McDonald's mentioned in the weather report time and again?

2. An Introduction to the U. S. Climate

(1) Micro-listening

　　Because the United States is so large, it has many different types of climate. The southern part of the West Coast has what is called Californian climate. It is warm and sunny most of the year. The Northern Pacific Coast has a marine climate, with mild winters and relatively cool summers. The central plains have a continental climate. Summers are hot and winters are very cold.

　　The Southeast is subtropical. South Florida is warm and humid all year round. On the East Coast, the climate is generally but not always continental. New England and the New York area are noted for their cold winters and hot summers. The land near the Rocky Mountains has a highland climate. It is cool because of its high altitude and has some rain and snow.

　　Alaska has a cool marine climate, but its central portion has a cold continental one. Hawaii is warm and humid all the year round.

(2) Answer the following questions:

　　1) Why is the U. S. climate varied in types?

　　2) What is the climate like on the Northern Pacific Coast?

　　3) Are summer and winter temperatures the same or different in the New York area?

　　4) What is the difference between summer and winter on the East Coast and the West Coast?

　　5) Which part of the U. S. has a highland climate?

　　6) What part of the United States is called New England? What is the climate like there?

3. A U. S. Synoptic Chart

(1) Micro-listening

Weatherman: Good morning, ladies and gentlemen! The national weather map shows a high-pressure area all along the eastern coastline, which has brought them very pleasant sunny weather from New York to Florida. But showers and thunderstorms are occurring from the Ohio River all the way south to the Gulf Coast, depositing heavy amounts of rain over the southern states.

By far the worst of the storms have occurred in and around Texas. Some Texas stations have reported up to 15 inches of rain in a twenty-four-hour period with high winds and thunderstorms. Two tornadoes were reported along the Gulf Coast in Texas, but we have no confirmation of damage or injury.

Heavy amounts of snow were reported in the Rocky Mountain region with record cold temperatures in Denver and Boulder. As much as a foot of snow has fallen in some of the mountain stations.

In contrast, temperatures in Arizona and the desert southwest went over the one-hundred-degree mark again today under bright, sunny skies. And that winds up our

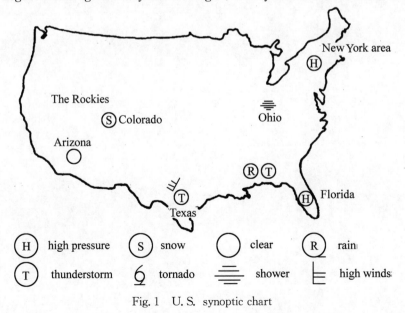

Fig. 1 U. S. synoptic chart

weather report for this morning.

(2) Answer the following questions:

1) Where is the high-pressure area for today on the U. S. map?
2) Is today's weather on the East Coast sunny or cloudy?
3) Besides the storm in Texas, where else is there heavy rain and thunderstorms?
4) How much rain was reported in Texas?
5) Where were tornadoes reported?
6) What does the phrase "record cold temperatures" mean?
7) How heavy is the snow which has fallen in some places of the Rocky Mountains?
8) What is the weather like in the desert southwest of the United States?

4. An Area Weather Map

(1) Micro-listening

On the area weather map most stations in southern Michigan are still reporting sunny skies. It's seventy-nine degrees at Detroit, and seventy-three degrees at Lansing. Chicago is reporting light showers. South Bend is cloudy as the clouds move in from the southwest.

The temperature at Ann Arbor Airport in degrees Celsius is twenty-three point three. That's seventy-four degrees on the Fahrenheit scale. Sixty-six degrees is the water temperature of the lake with winds gusting at twenty knots. The relative humidity is fifty-five percent and the barometric pressure is thirty point eleven inches of mercury and falling.

The pollution index today is seventy-five. The quality of our air is fair. Sunrise will be at six o'clock tomorrow morning.

And now for the extended forecast. For tonight, we expect partly cloudy conditions and mild temperatures with tonight's low about sixty degrees and only a twenty percent chance of any showers this evening.

Tomorrow morning looks for mostly cloudy conditions with a seventy percent chance of showers and thunderstorms continuing into the evening. It looks as if it will be a mild, but rainy weekend.

(2) Answer the following questions:

1) What's the weather like in Southern Michigan?
2) Why is South Bend cloudy?
3) What is the air temperature at Ann Arbor in degrees Celsius?
4) What is the pollution index?
5) At what time will the sun rise tomorrow?
6) What will tomorrow evening's weather be like?

5. Weather Forecasting in the U. S.

(1) Micro-listening

Accurate weather forecasting depends on the forecaster's knowing as much as possible about the total state of the atmosphere. The National Weather Service operates a far-flung network of stations where such information is collected and speedily sent to regional forecasting centers. Observations are taken hourly, day and night, at some 400 stations. More complete observations are made every 6 hours at many of the stations. Upper air elements are sampled by radiosonde twice daily at some 50 stations. Pilot balloons, measuring the upper air winds, are sent up four times a day at some 125 stations. The Coast Guard, Air Force, Navy and Federal Aviation Administration cooperate in observing and forecasting the weather. Over 9000 part-time weather stations also report to the National Weather Service.

Weather affecting the United States a week or more from now is being born today in air masses over other countries and the oceans. A world-wide network of weather stations (under the World Meteorological Organization, an agency of the United Nations) exchanges data with the National Weather Service. Merchant ships radio information four to eight times a day while at sea, and airliners report weather conditions encountered in flight. Because of all this, the National Weather Service is able to achieve better than 85 percent accuracy in its 24-hour to 36-hour forecasts.

(2) Answer the following questions:

1) On what does accurate weather forecasting depend?
2) Who else cooperates in observing and forecasting the weather?
3) With whom does the N. W. S. exchange its weather data?
4) Do airliners make any contribution to collecting weather data?
5) Is there any world-wide cooperation in making weather forecasts?

6. Meteorologists and Their Work

(1) Micro-listening

There is an old Chinese saying, which goes "A storm may arise from a clear day; something unexpected may happen any time." That is to say, things in nature are unpredictable. This is not really true today. Something is indeed being done. Today meteorologists try to make people's lives safer and better. Meteorologists are constantly studying the weather. Some meteorologists observe the weather. Others analyze the weather information and still others make forecasts about the weather. Many forecasts help to warn people of approaching storms and other bad weather.

The United States Weather Service operates a network of weather stations throughout the U. S. The Weather Service has more than 400 stations collecting and recording weather information. At these stations weather observations are taken every hour both during the day and at night. The Weather Service issues 24-hour weather forecasts. It also issues five-day forecasts and even thirty-day forecasts. It may seem hard to believe that some types of weather forecasts are 95% accurate.

In 1959 the United States launched its first weather satellite. This satellite was specifically designed to collect, record and send back weather information. Since that time, several weather satellites have been sent into space. They continue to provide valuable weather information to meteorologists in all parts of the world. Accurate weather forecasts can save thousands of lives and millions of dollars in property damage. Meteorologists are, indeed, helping to make our lives better and more secure.

(2) Answer the following questions:

1) How many weather stations does the U. S. Weather Service operate?
2) What kinds of weather forecasts does the Weather Service issue?
3) How accurate are some weather forecasts?
4) When did the United States launch its first weather satellite?
5) What do weather satellites do?
6) What can timely and accurate weather forecasts contribute to people?

7. To Rain or Not to Rain

(1) Micro-listening

In the past, Chinese weather reports would state whether or not it was going to rain that day by announcing that the day would either "have precipitation" or "not have precipitation". Now, however, we have to judge ourselves whether or not it will rain by listening to the probability of precipitation in the weather forecast.

Many people do not understand what the probability of precipitation means and why the Central Meteorological Observatory has changed the way they report the weather. Some people even criticize the meteorological agents for escaping their liability. In fact the probability of precipitation, which took the place of the simple but subjective weather prediction of the past, is more scientific.

The new weather forecast is not the only new thing. As the country opens to the outside and introduces reforms, Chinese are exposed to many new "strange things". A survey showed that about 60 percent of urban residents had no idea what taxable income was, though China started to collect personal income tax in 1980. New things such as probability of precipitation and income tax represent, to some extent, the development of the society.

But understanding these new things takes time and strength to adapt to change. The public in a modern society has to become more open-minded and be ready psychologically for the changes. Blindly sticking to convention and conservativeness will unavoidably lead to stagnation and backwardness.

(2) Answer the following questions:

1) What change has been made in the weather forecast?
2) How do many people look at the new way of weather reporting?
3) Do people like the new way of weather forecast?
4) What is the speaker's attitude towards the new way of weather forecast?
5) What does the speaker mean by mentioning income tax here?

8. Hail

(1) Micro-listening

Hail is precipitation in the form of hard rounded pellets or irregular lumps of ice. Usually, hailstones have a diameter of about 1 centimeter, but they may vary in size from 5 millimeters to more than 10 centimeters in diameter. The largest hailstone on record fell on Coffeyville, Kansas, September 3, 1970. With a diameter of 14 centimeters and a circumference of 44 centimeters, this "Giant" weighed 766 grams! The destructive effects of heavy hail are well known, especially to farmers whose crops can be devastated in a few short minutes and to people whose windows are shattered.

Hail is produced only in cumulonimbus clouds where updrafts are strong and where there is an abundant supply of supercooled water. Hailstones begin as small embryonic ice pellets that grow by collecting supercooled cloud droplets as they fall through the cloud. If they encounter a strong updraft, they may be carried upward again and begin the downward journey anew. Each trip through the supercooled portion of the cloud may be represented by an additional layer of ice. Hailstones may also form from a single descent through an updraft.

In either case hailstones grow by the addition of supercooled water on growing ice pellets. The ultimate size of the hailstone depends primarily on three factors: (i) the strength of the updrafts, (ii) the concentration of supercooled water and (iii) the length of the path through the cloud.

(2) Answer the following questions:

1) How big are hailstones usually?
2) How big was the largest hailstone on record?
3) What damage can be caused by hail each year?
4) What are embryonic ice pellets?
5) How are hailstones produced?
6) How does the largest hailstone form?

9. Is Snow Coming?

(1) Micro-listening

The weather was wet and dreary across the Midland Thursday. In Missouri up to 7 inches of rain has fallen in two days. A flood warning was posted.

A Kansas woman who was jogging during a sudden storm was struck by lightning and killed. Her dog, who was running beside her, was not hurt.

It was snowing in Michigan. Up to four inches of snow was expected in New York State and Vermont.

Hailstones 1.5 inches around hit Holdrege, Nebraska and golf-ball-sized hail hit Newport, Arkansas. Pea soup fog spread through the Missouri Valley and into Kentucky and Indiana. The rain in the Midwest was the result of a typical weather pattern. It came up from the Gulf of Mexico.

Thunderstorms rumbled across parts of Texas, Nebraska, Illinois, South Dakota and Florida.

Meanwhile, heavy rain is expected to cause flooding in Portland, Oregon.

(2) Answer the following questions:

1) What did the weather report say in Missouri?
2) What happened to a Kansas woman and her dog?
3) Was it snowing in Michigan?
4) What happened in Newport and the Missouri Valley?
5) What was the weather like in the Midwest?

10. Typhoons

(1) Micro-listening

Typhoons and hurricanes are nature's most destructive force. These storms have taken millions of lives——one storm in Bangladesh killed 300000 people in 1970—— and caused billions of dollars worth of damage.

Typhoons occur in the Philippines or China's seas and hurricanes usually take place in the North Atlantic Tropics or the Western Pacific. However, people who live in the Indian Ocean call them cyclones. Whatever name is used, a typhoon is a terrible windstorm that includes heavy rains, high waves and high tides. A typhoon is hundreds of miles across, with winds that whirl around in a great circle at speeds at least 75 miles an hour. The winds may reach speeds of 200 miles an hour but about 100 miles an hour is more common.

Typhoons form over tropical oceans not far from the equator, where the air is very moist. Certain weather conditions such as a thunderstorm may cause great quantities of moisture to condense. In other words, vapor is turned to droplets of water. The condensing moisture gives up large amounts of heat, which warms the air. The warmed air rises rapidly, and as the earth rotates the column of warm air begins to spin, forming a typhoon. In the Northern Hemisphere, a typhoon or a hurricane spins anticlockwise, while in the Southern Hemisphere it spins clockwise.

Usually a typhoon lasts several days. Generally speaking, a newly formed typhoon moves westward and then turns away from the equator. At first a typhoon moves forward at speeds of about ten miles an hour. But after the storm turns away from the equator, its speed increases to 30 or 40 miles an hour. As the storm moves, the moisture falls as heavy rains.

At the centre of the typhoon is a column of air only a few miles in diametre that spins very slowly. This is the eye of the typhoon. In the eye, there is little wind or cloudiness. In fact the sun may even shine through the typhoon's eye! After the eye of the typhoon passes over an area, the storm brings more rain and high winds. When a typhoon moves off the ocean and travels inland, friction between the rough land surface and the winds finally slows the storm down.

(2) **Answer the following questions:**

1) What disaster did the storm in Bangladesh bring to the people in 1970?
2) Why does this terrible windstorm have three different names?
3) Where do typhoons come from? Why?
4) How do typhoons form?
5) Does a typhoon or a hurricane spin clockwise or anticlockwise?
6) How fast does a typhoon move forward?
7) What is the eye of a typhoon? What is it like?
8) Why do the strong winds finally slow down?

11. Hurricanes

(1) Micro-listening

The eye of a hurricane is at the center of the storm——a zone of near calm or light breezes, with clear or lightly clouded skies overhead. It averages about 20 miles in diameter. Thousands, ignorant of hurricane's anatomy, have gone out into the calm of the eye, unaware that they would soon be hit again by the full might of the other side of the hurricane.

The eye may be caused by centrifugal force acting on winds at the rim of the eye. The centrifugal force acting on a rotating body doubles when the radius of rotation is cut in half. As air spirals in toward the center of a hurricane its centrifugal force increases greatly. The cloudy wall of the eye is where the centrifugal force exactly balances the pressure, forcing air inward to the low-pressure center. Friction with the ocean surface slows down the whirling air and decreases the centrifugal force, So the eye is small at the surface. Aloft where wind speeds are great, the centrifugal force is higher and the eye larger and funnel-shaped.

The life history of a hurricane can be traced from its birth as a tropical low, through maturity, and to decay, as an extratropical cyclone in the westerlies. The map below shows a storm starting west of the Cape Verde Islands as a tropical low with winds less than 32 mile/h. Next we find tropical storm "Betty" (hurricanes are given women's names) with winds 32 to 73 mile/h, approaching the Leeward Islands of the West Indies. When next reported it is a mature hurricane with winds over 75 mile/h right over Guadeloupe. It continues westward at 10 to 12 mile/h, passing south of Puerto Rico and Haiti. Then it begins to curve north, still at 10 to 12 mile/h, passing over Cuba and the Bahamas, but now with winds over 125 mile/h. Betty then curves to the northeast moving at 30 mile/h until it hits a cold front north of Bermuda. Hurricane Betty induces a wave on the front and becomes an extratropical low, ending as a storm over the North Sea. Hurricane Betty's path is typical. It could have continued west to hit the Gulf Coast Florida or the Atlantic seaboard.

(2) Answer the following questions:

1) What is the eye of a hurricane?

2) How wide is a hurricane's eye usually?
3) Why is the eye smaller at the surface?
4) Is Betty's path unusual?
5) Could Florida be hit by Betty?

12. El Nino Phenomenon

(1) Micro-listening

A massive pool of warm ocean water in the South Pacific is causing changes in the atmosphere, according to the Climate Centre in the U. S. The warm water could produce unusual weather around the world in the next few months. As a result of this phenomenon, known as El Nino, more rainfall than normal is likely this winter across some areas of the United States, with unusually, warm or cold weather in other parts of the country.

Currently the phenomenon is marked by a warm pool of water along the equator. The warm water covers such a large area that it extends from the International Date Line to the coast of South America. That water is nearly 4 degrees Fahrenheit above normal.

As this warm water spreads almost a quarter of the way around the globe, it has a global influence, especially on the weather. The reason is that it disturbs the atmosphere. El Nino is probably the most important climate event beyond the annual cycle of seasons. Because the changes seemed to be first noticed around Christmas time, the phenomenon was given the name El Nino which is Spanish for child, a term often used to refer to Baby Jesus.

El Nino occurs every three to five years, sometimes in a mild form and sometimes strongly affecting weather patterns worldwide. Details of its causes are not yet fully understood by meteorologists. When El Nino occurs, unusually warm air can be pumped into Canada, Alaska and the northern United States. At the same time, conditions tend to be wetter than normal along the U. S. Southeast Coast and the Gulf of Mexico. Also, the Atlantic and Caribbean hurricane season tends to be milder than usual.

(2) Answer the following questions:

1) What is the phenomenon of nature, known as El Nino?
2) Where does the word El Nino derive from?
3) What happens to the global weather pattern during El Nino?
4) Is El Nino just the same climate event as the annual cycle of seasons?
5) How often does El Nino occur?
6) Do scientists fully understand its causes?

13. The Greenhouse Effect

(1) Micro-listening

In the atmosphere, carbon dioxide acts rather like the glass in the roof of a greenhouse which allows the sun's rays to enter but prevents the heat from escaping.

According to the prediction of a weather expert, the atmosphere will be three degrees Centigrade warmer in the year 2050 than it is today, if man continues to burn fuels, such as coal and oil, at the present rate. If this warming-up were to take place, the ice caps in the poles, that is, the north and south poles of the globe, would begin to melt. As a result, the sea level would rise several metres and coastal cities would be severely flooded. Also the increase in atmospheric temperature would lead to great changes in the climate of the earth, the northern hemisphere in particular.

In the past, attention was mainly paid to a man-made warming of the earth in the Arctic region. It is because the Antarctic is much colder and has a much thicker ice sheet. But the weather experts are paying more attention to the west Antarctic as this area may be affected by only a few degrees of warming. That is to say, the temperature in the west Antarctic will possibly increase in the next fifty years from the burning of fossil fuels.

Satellite pictures show that large areas of Antarctic ice are already disappearing. The evidence available shows that a warming has taken place. Meteorologists say that it is carbon dioxide in our atmosphere that warms the earth. Meteorologists call this phenomenon the greenhouse effect.

However, scientists say up to now natural influences on the weather have been greater than those caused by man. The question is: which natural cause has the most effect on the weather?

One possibility is the activity of the sun. Sun spots seem to have a considerable effect on the distribution of the earth's atmospheric pressure and consequently on wind circulation. Another possibility is volcanic eruptions. Explosive volcanic eruptions may cause changes in the earth's climate because the long-lived volcanic clouds will absorb radiation which, in turn, will lower air temperature.

(2) Answer the following questions:

1) What is CO_2 in our atmosphere compared to?
2) What does it mean by "the greenhouse effect"?
3) What would happen to the earth if global warming took place?
4) Which region did scientists pay more attention to in the past, the Arctic or the Antarctic?
5) What is the evidence that satellite pictures show to us?
6) What effect do the sun spots have on the earth?
7) Why is it that volcanic clouds may cause changes in the earth's climate?
8) Who do you think is to blame for this greenhouse effect, man or nature?

14. The Threat of Global Warming

(1) Micro-listening

Studies show that temperatures in 1990 will be higher than any year since people began keeping records of temperatures. Scientists say the record warmth last year continued a condition first observed in the 1980s. Average temperatures on earth have increased in the last 111 years. But the greatest increase has been in the past eleven years. Most of the warmest years since 1880 have occurred since 1980. Some scientists say these findings strengthened the possibility that the greenhouse effect has begun. Greenhouses are clear-glass or plastic buildings used to grow plants. The glass or plastic traps the heat of the sun. Even in bitter cold weather the plants inside survive. The earth's atmosphere acts like a greenhouse. Carbon dioxide and other gases in the atmosphere trap the heat from the sun. They prevent the heat from escaping into outer space. This balanced system makes it possible for plants, animals and people to survive on earth.

Now, however, the balance is changing. Human activities are producing greatly increased amounts of carbon dioxide and other gases. And the gases are trapping more heat. Scientists who believe in the greenhouse effect say increased temperatures will have major effects on the earth's environment, agriculture and sea levels. They are testing their idea all the time. American and British scientists organized the latest studies. Each group examined the records of temperatures on land. The British scientists also examined records of temperatures at sea. The American scientists say the average temperature last year was 15 and 56 hundredth degrees Celsius. That is the highest average temperature since their records began in 1880. It is almost one tenth of a degree higher than the old record high.

The American scientists say parts of the eastern United States, Europe and Asia had higher than normal temperatures. Only Greenland and parts of eastern Canada had lower than normal temperatures. The British records go back to 1850. The British scientists say they believe 1990 was the warmest year since then. They found evidence of record high temperatures in Europe, western Siberia, East Asia and most of North America.

(2) **Answer the following questions:**

1) What do studies show with regard to the earth's temperature?
2) When did most of the warmest years take place since 1980?
3) Why can the plants inside a greenhouse survive even in bitter cold weather?
4) What role do carbon dioxide and other gases play in the earth's atmosphere?
5) What are the major effects of greenhouse gases on man and the earth?
6) What latest studies have the American and British scientists organized?
7) What are the scientific findings drawn by the American scientists?
8) Why do the British scientists believe that 1990 was the warmest year since 1850?

15. Ozone

(1) Micro-listening

The gas now called ozone has been known for more than two centuries by the odor experienced in the presence of all kinds of electrical discharges. The odor was originally thought to be the smell of electricity until in 1840 Schoenbein showed that ozone was a separate chemical compound, now known to be a triatomic oxygen allotrope.

Ozone is an effective disinfectant. It is used to purify some municipal water supplies. Early students of ozone, who were aware of this bactericidal action also discovered that ozone was absent in smoky rooms, mines and industrial areas and surmised a connection between ozone and good health.

A number of ironics arise with regard to ozone. Originally, its presence in the troposphere was taken as a sign of good health; now it is a criterion of air pollution, and national standards limit allowable exposure in several countries. Nevertheless, its presence in the stratosphere is essential to life on earth, and environmental threats to stratospheric ozone are themselves subject to regulation. High-flying aircraft emitting oxides of nitrogen in the stratosphere have been suspected of destroying the ozone layer. If the same aircraft fly into Los Angeles, atmospheric chemistry acts on their hydrocarbon and oxides of nitrogen emissions, and they are suspected of producing ozone, together with the other undesirable products of photochemical smog. It is ironic that the same aircraft could destroy ozone in the stratosphere where it is essential to life on earth and produce ozone in photochemical smog where it is regarded as a potential health hazard.

(2) Answer the following questions:

1) What was ozone thought to be before and what is it known as now?

2) What did early students of ozone discover? What conclusion did they come to with it?

3) What was the presence of ozone taken as in the troposphere originally?

4) At which layer of atmosphere is ozone essential to life on earth?

5) Why do we say environmental threats to stratospheric ozone are themselves subject to regulation?

16. The Hole over the Antarctic

(1) Micro-listening

Atoms of oxygen usually link themselves together in pairs. They form the oxygen molecule (O_2). However, a few oxygen atoms join together in three. They form ozone (O_3). Such chemical processes are taking place in the stratosphere, which is about 15—50 kilometres above the earth's surface.

Ozone is not stable because a grouping of three is not the preferred arrangement for oxygen atoms. It makes ozone very reactive because it tries to form stable compounds with other chemicals.

Although ozone is a colourless, poisonous gas, the layer it forms helps to protect all life on earth from ultraviolet radiation (UVR). The full blast of the sun's ultraviolet light is biologically harmful in many ways. If this ozone were lost, more ultraviolet light would reach the earth. The strong ultraviolet light would burn our skins, blind our eyes, and eventually result in our destruction.

But the ozone layer is only paper-thin at best. Studies have shown that the ozone layer has been thinning even more in recent years. Not only is the ozone layer thinning, but it has disappeared entirely in some places, on a temporary basis. A "hole" in the ozone layer has developed over Antarctic every year since 1979, and duration of the missing layer has increased every year. In 1988 a second ozone hole was found over the Arctic for the first time.

After years of study and observation, scientists have found the answer to the so-called "ozone hole". They believe that the ozone layer is being destroyed by certain man-made chemicals known as chlorofluorocarbons, or CFCs. CFCs are widely used in refrigeration and air-conditioning, in foam and plastic manufacturing and in spray cans. Although CFCs are extremely stable in the lower atmosphere, at high levels CFCs are broken down by ultraviolet radiation. When this chemical reaction occurs, chlorine atoms are produced. Now the free chlorine atoms will react with ozone in the upper levels of the atmosphere. It has been proved that as many as 100000 ozone molecules can be removed from the atmosphere for every chlorine molecule released. Since ozone is destroyed faster than it forms, no wonder the ozone hole has appeared.

Serious steps are now under way to ban the use of CFCs. One hopes that it is not too late.

(2) **Answer the following questions:**

1) How many oxygen atoms does it take to make an ozone molecule?
2) How far is the ozone layer from the earth' surface?
3) Why is the ozone layer protective? What is its width?
4) What harmful effects may ultraviolet radiation cause?
5) Where did the scientists find the ozone hole?
6) What do CFCs refer to?
7) How can man prevent the ozone layer from being destroyed?

17. How Solar Activity Affects Life on Earth

(1) Micro-listening

For months scientists around the world have observed the increased activity on the sun. The activity is in the form of sun spots, geomagnetic storms and solar flares. A group of scientists in the United States is watching the sun very closely now. The scientists work at the Space Environment Laboratory in Boulder, Colorado.

The laboratory is part of America's National Oceanic and Atmospheric Administration. Even a small increase in solar activity affects life on earth. For example, it is known to damage satellites. It pushes a satellite from its orbit back toward the earth. This can be a serious problem if the satellite provides information to ships and planes. Increased solar activity also affects some defense warning systems. It increases electronic noise level and creates false signals. In short-wave radio broadcasting, distant stations become more difficult to hear.

It is the job of the scientists at the Space Environment Laboratory to study solar activity and give advice about its effects. They provide this advice to other scientists, governments and anyone else who is interested. There are different ways to measure solar activity. One method is to count sun spots. These are areas of the sun that are "cooler" than surrounding gases. They appear darker than surrounding areas. Another way is to measure geomagnetic storms and exploding solar flares. Such flares shoot out radiation and millions of tons of matter into space.

The scientists at Colorado get much of their information about the sun from telescope pictures. They use telescopes in Australia, Italy, Puerto Rico and the Untied States. They get other information from weather satellites. The Space Environment Laboratory also exchanges information from scientific centres in other countries. It communicates with scientists in Australia, Canada, France, Japan, Poland and the former Soviet Union.

All this information is put into a computer. The computer organizes it, so scientists better understand what is happening on the sun. America's Space Environment Laboratory in Boulder, Colorado is perhaps best-known for messages broadcast by Radio Station WWV. These messages tell what is happening on the sun right now and for the next several hours.

(2) Answer the following questions:

1) What have scientists around the world been observing for months?
2) What activities are going on in the sun?
3) Why do sun spots appear darker than surrounding areas?
4) What effect do sun spots have on our life on earth?
5) How do scientists measure solar activity?
6) What do geomagnetic storms refer to?
7) Who does the Space Environment Laboratory exchange information with?
8) Where do scientists keep all their information about the activity of the sun?

第三部分　气象科技英语知识

Part Three: Knowledge of English for Meteorological Science and Technology

第三部分　气象科技英语知识

1. 科技英语的特点

科技英语(English for Science and Technology,简称 EST)是专用英语(English for Specific Purposes,简称 ESP)的一种,它是用来描述科技用语中各种语言现象和特性的一种英语体系。科技英语是用来进行国际交流的重要手段,据统计,当今世界上约有三万种英文科技刊物,约有 75% 的科技资料是英文发表的,在口头交流时也普遍使用英语,所以科技英语教学受到普遍重视。

科技英语作为一种英语语体,和普遍英语相比有其一定的独特之处。一般来说,它有下列特点。

(1)文体方面,通常有结构严谨、逻辑性强、形式多样、类型复杂(有论文、文摘、实验报告、专题评述、专利说明等)等特点。

(2)语法方面,常用被动语态和倒装句等,例如:

Heat can be converted to energy.

The barometric principle was invented by Torricelli in 1593.

Atmospheric pressure decreases with increase in altitude and so does the density of the atmosphere.

上面前两句用的是被动语态,第三句的后半句是倒装句。

(3)句型方面,句子长,长句多(后文将举例分析)。

(4)词汇方面,有很多专业词汇,而且随着学科发展不断有新词汇(或词组)出现。据统计,在科技英语文献中专业词汇约占总词汇的 20% 以上。

(5)文章方面,需要具有一定的专业知识才能正确地理解词意和文意。

总的来说,科技英语包括四个基本要素,即词汇、语法、习惯用语和专业知识,四者相互联系,不可分割。

2. 科技英语的词汇

2.1　专业词汇与普通词汇

　　科技英语中有不少专业词汇，它们多数是单义词，而且一般只有字面意义，相对来说较易掌握。科技英语中有更多普通词汇，它们常常一词多义，而且会引起联想意义，加上许多词汇用法灵活，所以必须根据上下文及熟悉其习惯用法才能正确地判断其意义。例如 look 与 at，up，after，for，into 等不同介词连用，便有不同意义。因此，在科技英语中，我们不仅要注意专业词汇，而且也要注意普通词汇，尤其是多义，多功能的普通词汇。

2.2　生词的记忆

　　词汇量大小是英语水平的标志之一。生词要靠反复记忆才能掌握。但也可利用记忆法来帮助记忆。例如可用"逻辑记忆法"，即将某些词看作是常用词的变形，便较容易记住。例如"comet（彗星）"可看作为"come t"，这样便容易记住其拼写。也可根据构词法来掌握生词。

2.3　构词法

　　科技英语中的专业词汇，很多都按一定的构词规律构成。因此，学习构词法对掌握生词和扩大词汇量是很有帮助的。
　　科技英语构词法主要有转化（如 an empty box 和 to empty a box 中的 empty 分别有"空的"和"腾空"意义，词性分别为形容词和动词），合成（如：frontogenesis——锋生，cyclogenesis——气旋生），以及派生等规律。
　　所谓派生构词法就是指在一个词干上加前缀（词头）或后缀（词尾）来构成新词的方法。如 nonlinear（非线性的），cloudless（无云的）等。所以掌握前后缀的用法有助于扩大词汇量。
　　下面列举的是气象科技英语中常见的前缀和后缀，以及它们的附加意义和例词。作为练习，你可在补遗栏中添上更多的例词。

气象科技英语中常见的前缀和后缀,以及它们的附加意义和例词

前(后)缀	附加意义	例 词		补 遗	
un-	否定	unstable unequal	不稳定的 不相等的	unsatisfactory	不充分的
in-	否定	inaccurate invisible incompressible infinitesimal	不精确的 不可见的 不可压的 无穷小的	instable inhomogeneous independent informal indefinite	不稳定的 不均匀的 独立的 非正式的 不定的
ir-	否定	irregular	不规则的		
im-	否定	impossible imbalance immature	不可能的 不平衡的 不成熟的		
il-	否定	illimitable illogical	无限的 不合逻辑的		
non-	否定	non-homogeneous nonconvective nondivergent	非均匀的 非对流性的 无辐散的		
dis-	否定	discontinuity	不连续性	dissociate	分解
de-	否定	decelerate	减速		
a-	否定	ageostrophic adiabatic asymmetric aperiodic	非地转的 绝热的 非对称的 非周期性的		
co-	共同	co-existence cooperation	共处 合作	correlation	相关
anti-	反抗	anticyclone antibody	反气旋 抗体		
counter-	反对	counterclockwise	逆时针地		
contra-	反,相反	contrary	相反的		
inter-	相互	interface interaction international	界面 相互作用 国际的	interplanetary	星际的
pre-	预先	pre-set pre-heat	预置 预热	prediction	预报
sub-	亚 次级的 下级的	subtropical substorm subgrid subdivision subsystem	亚热带的 亚风暴 次网格 细分,再分 分系统		
trans-	横过	transcontinental	横贯大陆的		

前(后)缀	附加意义	例 词		补 遗	
super-	超过,在…上	supercooling superheat supernormal	过冷却 过热 超常	superadiabatic	超绝热的
over-	超过	overdevelop	过度发展		
ultra-	超	ultraviolet ultra-long wave	紫外线 超长波		
under-	下,不足	underdevelop	发展不足		
semi-	半	semi-arid semi-permanent	半干旱 半永久		
hemi-	半	hemisphere hemicycle	半球 半周期		
multi-	多	multicell	多单体		
poly-	多	polygon polynomial	多边形 多项式		
tele-	远	telescope telemetry	望远镜 遥测学		
centi-	百分之一	centigrade centimeter	百分度的,摄氏度 厘米		
milli-	千分之一	millimeter millibar	毫米 毫巴		
micro-	微	microscope microclimate	显微镜 微气候		
meso-	中等的	mesoscale meso-cyclone	中尺度 中气旋	mesopause mesosphere	中间层顶 中间层
macro-	大的	macro-climate	大气候		
kilo-	千	kilogram	千克		
post-	在…之后	post-processing post-frontal	后处理 锋后的		
extra-	在外的 超额的	extratropical extra-long wave extra long-range	温带的 超长波 超长期		
iso-	等	isobar isotherm	等压线 等温线		
pseudo-	假	pseudo-adiabatic	假绝热		
quasi-	准	quasi-static quasi-geostrophic	准静力的 准地转的		
cumulo-	积状	cumulonimbus	积雨云		
strato-	层状	stratocumulus	层积云		

前(后)缀	附加意义	例 词		补 遗	
cirro-	卷状	cirrocumulus	卷积云		
geo-	地	geostrophic geology	地转的 地理学		
bio-	生物	biosphere	生物圈		
topo-	地形	topography	地形学		
baro-	气压的	baroclinic	斜压的	barotropic barometer	正压的 气压表
hydro-	流体水	hydrostatic	流体静力的		
homo-	等，均	homogeneous	均匀的		
bi-	双	biweekly biennial	双周 两年的		
tri-	三	triangular	三角		
photo	光	photosynthesis	光合作用		
mono-	单	monochromatic	单色的		
isallo-	等变	isallobar	等高压		
-er -or -ist	构成名词表示行为者	forecaster predictor meteorologist	预报员 预报因子 气象学家		
-ion	构成名词表示行为性质、状态等抽象概念	observation radiation precipitation pollution evaporation dissipation condensation deflection fluctuation prediction separation	观测 辐射 降水 污染 蒸发 消散 凝结 偏转 振动 预报 分离		
-ing		cooling	冷却		
-ance -ence		resistance divergence convergence	阻力 辐散 辐合		
-ment		movement measurement	运动 测量		
-ure		pressure mixture temperature	气压 混合 温度		
-ics		hydrodynamics geophysics	流体动力学 地球物理学		

前(后)缀	附加意义	例　词		补　遗
-ness		usefulness stormness cloudness	有用性 风暴度 云量	
-ship		relationship	关系	
-ism		mechanism	机制	
-ity		density humidity	密度 湿度	
-y		difficulty efficiency	困难 效率	
-th		length width depth	长度 宽度 深度	
-logy		meteorology biology	气象学 生物学	
-ic(al)	构成形容词，表示具有某种性质、特征	meteorologic(al) climatic(al)	气象学的 气候的	
-ful		useful	有用的	
-ous		continuous enormous advantageous	连续的 巨大的 有利的	
-able		considerable comparable	可观的 可比较的	
-ible		compressible	可压缩的	
-ive		excessive quantitative	过量的 定量的	
-ent		different independent	不同的 独立的	
-ant		constant significant	定常的 重要的	
-an		American	美国的	
-ar		solar lunar	太阳的 月亮的	
-ary		arbitrary	任意的	
-less		cloudless colourless	无云的 无色的	
-ly	构成副词	extremely	极端的	

前(后)缀	附加意义	例　　词		补　　遗
-ward		backward upward poleward equatorward southward	向后 向上 朝极地 朝赤道 朝南	
-wise		clockwise	顺时针地	
-ize		visualize specialize miniaturize	形象化 专业化,专向化 使小型化	
-en	构成动词	lengthen shorten	使变长 使变短	
-fy		simplify	简化	
-ate		operate	操作	

合成词在科技英语中也十分常见。例如,upper-air(高空),ice-crystal(冰晶),thunderstorm(雷暴),hailstorm(雹暴),hailstone(雹块),duststorm(尘暴),nowcasting(临近预报),downburst(下击暴流),smog(烟雾 smoke＋fog)。最常见的合成词之一是用元音字母"O"或辅音字母"S"将某一音素组合起来而构成的词,例如 troposphere(对流层),tropopause(对流层顶),stratosphere(平流层),stratopause(平流层顶),hydrostatics(流体静力学),hydrodynamics(流体动力学),thermodynamics(热力学),等等。

3. 科技英语的语法

英语句子都是由词汇按一定规律组成的,这种语言结构的规律就是语法。学习语法有助于正确使用英语进行写作、说话,也有助于阅读和理解英语文献。在科技英语中经常会遇到长而难懂的句子,常常需要借助语法分析来理解它们。长句子常常是由于在主要成分上附加了许多修饰成分而构成的。我们只要对英语的基本句型非常熟悉,就不难找出长句中的"主要成分",然后再来处理"附加成分",问题就容易解决了。

3.1 英语的基本句型

英语句子是由一些基本成分构成的,这些基本成分包括主语(S)、谓语动词(V)、表语(P)、宾语(O)、补语(C)等。根据构成句子的基本成分的不同可以把英语句子划分成不同的句型。科技英语中常见的句型可分为一般句型和特殊句型两大类。其中一般句型又可分为主动句(谓语动词为主动语态)和被动句(谓语动词为被动语态)两类。下面是它们的基本格式:

一般主动句:

 (1) SV

 例:Air mass moves.

 (2) SVP

 例:Meteorology is the science of atmosphere.

 (3) SVO

 例:Computer assesses the observational data.

 (4) SV$O_{间接}O_{直接}$

 例:The sun gives us light.

 (5) SVOC

 例:We find it very useful.

一般被动句:

 (1) SV

 例:The balloon is released.

 (2) SVO

 例:A symbol is given each meteorological element.

 (3) SVC

 例:The condition is known as supercooling.

特殊句:

(1) 疑问句

例：Is oxygen a gas? （一般疑问句）

How (What, Where, When…) does radar work? （特殊疑问句）

(2) 命令句

例：Explain it, please.

(3) 存在句

例：There are many forms of energy.

(4) 省略句

例：T stands for temperature, P for pressure and V for wind speed.

(5) 强调句

例：Only thus can we do it better.

It was Galileo who invented thermometer.

It is the protons that are positive.

(6) 先行词 it 结构

例：It is necessary to…
                    that…

We deem it necessary to…
                      that…

(7) 比较结构

例：The more… The more…

The longer the better.

3.2 长句的分析

在基本成分上附加很多修饰成分就构成长句。一般来说，英语修饰成分有两种，即前置修饰和后置修饰。它们分别出现在被修饰成分之前或之后。例如，an important question（前置修饰），something important（后置修饰）。前置修饰一般容易分析，后置修饰的分析则较麻烦。例如：

The air which is moving swiftly usually has high kinetic energy.

The swiftly moving air usually has high kinetic energy.

上两句中，前一句就比后一句稍麻烦些。因此句子分析的重点不仅在于分析修饰成分，而且主要在于分析后置修饰成分。后置修饰成分除了个别由单词构成外，一般是由从句、短语构成的。能构成修饰成分的从句和短语有：介词短语、不定式短语、现在分词和过去分词短语、形容词从句（定语从句）、副词从句（状语从句）。

具体作句子分析时，可按下列步骤进行：

(1) 找出句子中的各种从句

从句一般可以凭借引导词，如 what, which, who, where, when, how, why 等来辨认。其中名词从句常用的引导词有 that, whether, if（以上为从属连词），who, which, what（以上为连接代词），when, where, how, why（以上为连接副词）。形容词从句常用的引导词有 who, which, that（以上为关系代词），when, where, why（以上为关系副词）。副词从句都由从属连词引导。

(2) 找出句中（包括从句中）的短语（介词短语、分词短语、不定式等）

短语一般凭借介词如 of, in…等，不定式标志 to，以及分词词尾-ing 或-ed 或其他不规则形式的分词来判断。

(3) 然后，分辨这些从句或短语在句子中的作用。即判断它们是句子中的基本成分还是修饰成分。

一种分辨方法是将这个从句或短语从句子中删去，若严重损害句子的基本结构或意义，则说明它是基本成分，否则便是修饰成分。

(4) 将从句、短语纳回到句型中去。再根据逻辑或专业内容，语言或语法要求以及上、下文来判断它们所修饰的成分。对于后置修饰，可以通过"向前推断"的方法来判断。即对该修饰成分之前的第一个、第二个、第三个……词逐个地进行推断，直至找出它所修饰的词为止。

通过上述分析过程，一般来说全句的语法结构和意义便已清楚了。下面我们来看一个实例：

The discovery of the upper air wave as an even more important seat of energy than the associated low level cyclone, followed by the theoretical analyst of Rossby (based on Helmholtz's vorticity theorem) before World War II, and of such investigators as Charney, Eady, Fjortoft, Eliassen, Starr, Kuo, Lorenz and Phillips, after the war, has finally led to the establishment of a fairly consistent picture of the large-scale motions of the atmosphere, culminating in Phillips successful numerical solution for the main properties of the general circulation (1956).

在上面的长句中，主要成分是 The discovery has led to the establishment. 这是 SVO 结构的一般主动句。所有其他的介词短语、副词及分词从句都是修饰主语、谓语和宾语的。像这一类的长句在《读物选》的课文中经常可见。作为练习，大家可以自己将它们找出来，进行分析练习，这对提高长句的阅读理解能力是有好处的。

4. 科技英语的习惯用语

英语除遵从语法规律外,还常常受习惯用法的支配。习惯用法的含义很广,这里只讨论一些惯常的结构搭配关系。

4.1 某些惯常的语法结构

英语中有些动词要求动名词作宾语,而有些动词则要求不定式作宾语。常见的要求动名词作宾语的动词有:avoid(避免),can't help(不得不),deny(否认),detest(嫌恶),dislike(不喜欢),enjoy(享受),escape(逃避),excuse(原谅),fancy(幻想),finish(完成),imagine(想象),keep(on)(继续),mind(介意),postpone(延误),practice(实行),risk(冒险),stop(停止),suggest(建议)。

常用的要求不定式作宾语的动词有:attempt(企图),dare(敢于),decide(决定),endeavor(力图),except(期望),fail(未能),fear(怕),guarantee(保证),offer(提出),pretend(假装),promise(答应),refuse(拒绝),undertake(承担),want(要),wish(愿望)⋯help(帮助)(不定式可省略 to)。

有的动词能用动名词或不定式作宾语。如:commence(开始),regret(遗憾)(这些动词多数用动名词作宾语);又如:hate(憎恶),intend(打算),learn(学习),like(喜欢),love(喜爱),prefer(宁可),propose(建议),try(试图)(这些动词多数用不定式作宾语);又如:begin(开始),cease(停止),continue(继续),forget(忘记),need(需要),remember(记得),start(开始)(这些动词既常用动名词,也常用不定式作宾语)。

4.2 词的习惯搭配

动词和介词,形容词和介词,名词和介词,名词和动词,名词和形容词,以及动词和副词(指不与介词同形的副词)之间常有习惯的搭配关系。了解这种搭配关系对科技英语阅读十分重要。例如见下句:

The symbol "t" refers not only in physics, chemistry, meteorology, or other branches of science and technology but also in our daily life to "time"(符号"t"不仅在物理学、化学、气象学或在其他的科技分支中,而且在日常生活中都指"时间")。在句中,refer to 是一种习惯搭配关系,表示"指⋯为"的意思。如果我们熟悉这种搭配关系,就很容易读懂这个句子。所以熟记一些习惯搭配(词组)是非常重要的。掌握词组的多少及使用的熟练程度在一定程度上反映了学习者的英语水平。

下面所列举的是在科技英语中常用的词组:

above all	*adv. phr.*	尤其是,最重要的是
accompanied with	*v. phr.*	带有,兼有
according to	*v. phr.*	依照,按照
accustom to	*v. phr.*	惯于
act for	*v. phr.*	代理
act on	*v. phr.*	起作用,起反应,作用于
adequate to	*adj. phr.*	够(用),敷(用)
adjacent to		接近…
adverse to		反,不利于
after all	*adv. phr.*	终究,毕竟
a fraction of	*n. phr.*	一小部分
again and again	*adv. phr.*	再三,屡次
agree to	*v. phr.*	同意,承认,赞成(事)
agree with	*v. phr.*	适应,相似,赞成(人)
a handful of	*adj. phr.*	少数
ahead of	*prep.*	在…前边;比…进步;优于;胜于
aim at	*v. phr.*	瞄准(目标)
all about		处处,到处
all along	*adv. phr.*	连续
allied to		有密切之关系
all in all	*adv. phr.*	全部,一切,第一
all manner of	*adj. phr.*	各种,各类
all one		全然相同
all over	*adv. phr.*	完全,到处
all round	*adv. phr.*	到处
all the same	*adv. phr.*	完全一样
alongside of	*prep.*	在…的侧面;与…并排
along with	*adv. phr.*	与…一道,与…同时
and others (etc.)		等等
and so forth		等等
and so on		等等
and the like		等等
answer to		符合
anterior to		前面的,前部的

apart form	prep.	且不说,除开
approve of	v. phr.	承认,赞成
as a consequence	prep.	因此,从而
as against	adv. phr.	比(对)…
as a matter of fact	adv. phr.	实际上
as a result of	prep.	…的结果
as a rule	adv. phr.	大概,通常,照例
as far as	adv. phr.	至于,尽…所,(直)到
as follows	adv. phr.	如下
as for =as to	prep.	至于,就…而论,论及
aside from	prep.	除…外;且别说
as if =as though	conj.	恰像…一样
as long as	conj.	只要
as mentioned earlier		如前所述
as much	n. phr.	正如此,一样
as regards	prep.	关于
associated with	v. phr.	与…相联系
as soon as	conj.	一…,就…
as though		恰如,活像
as to	prep.	至于,就…而论,论到
as usual	adv. phr.	照常
as well	adv. phr.	并且,亦
as well as	conj.	和…一样;以及
at all	adv. phr.	全然
at all points		完全
at first	adv. phr.	首先
at full speed		以高速度
at last	adv. phr.	最后
at least	adv. phr.	少,至少
at present	adv. phr.	目前,现在
attach great importance to	v. phr.	重视
at the expense of	adv. phr.	牺牲
at the instance of		因…之请;因…的主张
at the most		至多,至极限

at the point of	prep.	差不多
at the top of	prep.	在…顶点
at variance with		异于,不合
back of	prep.	在…的背后,在…的后部
base on	v. phr.	以…为根据
bear in mind	v. phr.	记住
bear upon		靠,朝向,影响
because of	prep.	由于,基于
become of	v. phr.	结局,下场
before long	adv. phr.	不久,不多时
belong to	v. phr.	属于
below the mark		标准以下
made of		为…所制造;…所成
beyond doubt	adj. phr.	无疑地
beyond the bounds of		越出…区域之外
beyond the mark		超出界限,过度
beyond the reach of		为…力量不及
both… and …	conj.	既…又
bring about	v. phr.	引起,促成
bring on	v. phr.	开始,发生
but for	prep.	没有,若无…之助
but just	adv. phr.	仅仅,只不过
but that		除非,若不,除…之外
bound to		确实要,不得不…
build on	v. phr.	建筑…上
by accident	adv. phr.	偶然
by all means	adv. phr.	必然,务必
by and by	adv. phr.	逐渐地
by chance	adv. phr.	偶然
by dint of		因,用,由
by hand	adv. phr.	用手的
by inches	adv. phr.	渐渐
by oneself	adv. phr.	独自地
by main force		用强力

by means of		用,依靠,借…的力量
by no manner of means		决不
by no means		决不
by reason of	prep.	凭…的理由,因…,为了
by rule		按规则
by the aid of		借…的帮助
by the side of		在…的旁边
by virtue of	prep.	因为,凭,用
by way of	prep.	经由,通过,作为
cannot but	v. phr.	不得不
care for	v. phr.	注意,关心
carry away	v. phr.	带去
carry on	v. phr.	促进,处理
carry out	v. phr.	实行,完成
carry through	v. phr.	贯彻
cast up	v. phr.	计算
catch up with	v. phr.	赶上
center in	v. phr.	以…为中心
change into	v. phr.	变成
cite a case		举例
clear the land	v. phr.	驶离陆地
clinch an argument		定论,结论
close at hand		迫近,在即
close in upon	v. phr.	包围,围绕
close to	adj. phr.	接近
close to the wind		迎风
coincide with		一致,与…相符
combined in	v. phr.	化合成
come across	v. phr.	碰见
come after		在…之后
come away	v. phr.	离开
come forth	v. phr.	放出,提出,公布
come forward	v. phr.	前进
come in	v. phr.	进入,通行

come in its turn	v. phr.	挨次
come into	v. phr.	加入
come into force	v. phr.	实施,施行
come out	v. phr.	结果,结局
common to		共,通
communicate to	v. phr.	传达,传播
compare to	v. phr.	比拟
compare with	v. phr.	与…比较,对照
composed of	adj. phr.	由…而成
concerned in	adj. phr.	和…有关系;牵涉到
concerned with	adj. phr.	参与,干预
confine to	v. phr.	限;限制(于)
connect with	v. phr.	连接
consist in	v. phr.	在(于);存(于)
consist of	v. phr.	由…组成
contrary to	v. phr.	和…相反
contrast to	v. phr.	和…成对比
convert into	v. phr.	转化为
cool off	v. phr.	冷却
counter to	v. phr.	和…相反
cut out	v. phr.	省略
date from	v. phr.	从…起,始于…
deal out	v. phr.	分配,分给
deal with	v. phr.	处置,涉及
depend on	v. phr.	取决于,依靠
derived from	v. phr.	得自,由…得来
deviated from	v. phr.	脱离,违背
die away	v. phr.	渐停
differ from	v. phr.	不同于
differ in	adj. phr.	在…方面不同
distinguished from	v. phr.	区别于,与…有区别
differ on	adj. phr.	和…上不同
differ with	adj. phr.	和…不同;和…不合
different from	adj. phr.	和…不同

distinct from		明显的与…不同
divide into	v. phr.	分;分成
divided by		被…除
do with	v. phr.	有关系
do without	v. phr.	不用,舍去
draw conclusion		得出结论
draw near		临近
draw off	v. phr.	排除;取出
draw out	v. phr.	延长;抽取;分开
draw up	v. phr.	拉起;抽上;起草
drive at	v. phr.	企图;目的所在
drive to the wall		推开;轻视
due to	prep.	由于,因为
dwell on	v. phr.	详述
either …or …	conj.	或…或
enter into particulars	v. phr.	详述,细说
enter on	v. phr.	开始,着手
equal to	adj. phr.	等于
equip with	v. phr.	装备,设备
equivalent to	adj. phr.	同等的;等值的;等量的
ever and again		时时
even now		甚至现在还
even so	adv. phr.	即使如此
even though		虽然
ever since		自从
ever so	adv. phr.	非常,大大
every now and then		时常
every other		每隔一
every way		完全
except for		若无,除…外
except that		除…外
exclusive of		除去,不算
exert oneself to		努力
exposed to view		显而易见;出现

face to face	adj. phr.	面对面
fail to (inf.)		没有能
fall back on	v. phr.	凭借,依赖,投靠
fall into	v. phr.	分成,属于
fall off	v. phr.	减少,落,降
fall on	v. phr.	开始(行动)
fall out	v. phr.	发生
fall under	v. phr.	列入,归入
fall upon(on)	v. phr.	攻击,伤害
familiar with	adj. phr.	精通,熟悉
far and away	adv. phr.	非常,…得多
far and near	n. phr.	远近
far and wide	adv. phr.	到处
far away	adj. phr.	远,隔着
far from	adj. phr.	决无;决非
far off	adj. phr.	远隔的;远方的
fill in	v. phr.	填满
fill up	v. phr.	充满
filled with	v. phr.	充满
first and last	adv. phr.	大体上
fit for	adj. phr.	适于,适合
fit out	v. phr.	装备
fit up	v. phr.	准备,装备
fit with	v. phr.	装备,设备
for all that	prep.	虽然如此
for all the world		无论如何看;完全
for a space		片刻
for a time	adv. phr.	暂时
for certain		必定
for example	adv. phr.	例如
for good	adv. phr.	永远
for lack of	prep.	因为没有…的缘故
for fear of	prep.	唯恐
for instance		例如

for the good of		为…利益
for the ends of		为了…的目的
for the most part		通例,大概
for the present		当今
for the purpose of	prep.	为…起见,因要
for the sake of		为了
for the time being	adv. phr.	目下,现时
for the worse	adj. phr.	每况愈下,恶化
for want of		缺乏
free from(of)	adj. phr.	免于…,没有…
free…from	v. phr.	使…免于,使…从…解放出来
from age to age	prep.	一代一代
from among		从…之中
from hand to hand		传递
from the above mentioned		由以上所述
from the first		起初
from the point of view of		自…的观点
from the time of		从…以来
from within		从内部
gain upon	v. phr.	增进
get at	v. phr.	达到
get before	v. phr.	前进
get behind	v. phr.	落后
get down	v. phr.	下降
get familiar with		熟悉
get forward		上进
get off	v. phr.	离开,开始
get rid of	v. phr.	避免,摆脱
get the upper hand	v. phr.	操纵
get through	v. phr.	经过
get up	v. phr.	准备;研究;飞出;起
give effect to	v. phr.	实行;施行
give expression to		言,述
give rise to	v. phr.	产生

go about	v. phr.	着手,从事
go after	v. phr.	跟随
go halves		平分
go into	v. phr.	谈及
go into operation	v. phr.	实行,开始工作
go on	v. phr.	继续
go without	v. phr.	没有,缺少…
go by	v. phr.	过去,依照
have a care	v. phr.	注意
have in prospect	v. phr.	希望
have to		不得不
have to do with		有关系,关联
hand in hand	adv. phr.	连合,相随
hand over hand	adv. phr.	两手轮流行动
hardly any	pron. phr.	极少
heat up	v. phr.	加热
here and there	adv. phr.	处处;分散
hold back	v. phr.	阻止,不前进
hold good	v. phr.	有效,适用
hold in	v. phr.	压住,约束
hold on	v. phr.	坚持,继续
hold true		有效,适用
hold together		联合,结合,合成一体
in a broad (narrow) sense		广(狭)义来说
in accordance with	adv. phr.	依照,按照
in addition	adv. phr.	此外,再者
in addition to	prep.	除…之外
in advance of	prep.	在…之前
in a few words		简言之
in agreement with		和…一致
in a high degrees		非常,颇
in all cases		就一切情况而论
in all manner of ways		用各种方法,种种
in all (every-respect)		从各方面说

in a moment		立即
in any case	adv. phr.	无论如何,总之
in any event	adv. phr.	无论怎样都
in any sort	adv.	无论如何,必须
in any wise	adv.	无论如何
in a sense	adv.	在某种意义上
in as much as	adv.	因为,因…之故
in a word	adv.	总而言之
in behalf of	prep.	为…,为…的利益
in between		在中间
in case of	prep.	要是,如果,万一
in company with	adv. phr.	和…一道
in comparison with	adv. phr.	和…比起来
in compliance with		顺从,依
in common with	adv. phr.	与…相同
in conclusion	adv. phr.	最后,结论
in connection with	adj. phr.	关于,联合
in consequence of	prep.	因…之故,由于,因…之结果
in consideration of	prep.	因为,考虑到
in consistent with		和…不一致
in contrast with(to)		和…成对比的,和…大大不同
in contact with	adj. phr.	和…接触
in course of		在…中
independent of		不受…的影响
in detail		详细
in fact	adv. phr.	事实上
inferior to		劣于,次于,比…差
in front of	prep.	在…之前
in general	adj. phr.	一般来说
in keeping with	adj. phr.	与…相一致
in lieu of		代替
in light of	adj. phr.	根据,按照,借助
in line with	prep.	与…成直线,与…一致
in more detail		更详细

in nature	adv. phr.	性质上
in no time	adv. phr.	不多时
in opposition to		违反
in order to	conj.	为了…
in other words	adv. phr.	换言之
in particular	adv. phr.	特别
in place of	adj. phr.	代替
in point to		关于
in preference to		在…之先,胜(多)于…
in process of		在过程中
in proportion to	adv. phr.	按…的比例
in pursuance of		依…,按…
in question	adj. phr.	谈论中的
in reference to	prep.	关于
in regard to	prep.	关于
in relation to	prep.	关于,关系
in respect to(of)	prep.	就…而论
in return for	adv. phr.	作为…的报酬
in search of	prep.	寻求
in sight of	adj. phr.	可看见
in so doing		这样做时
in so far as		到…程度,在…范围之内
in spite of	prep.	不管,虽然
instead of		代替,而不
in support of		支持…,拥护…
in terms of	prep.	依…,据…,用…(字眼)表示
in the case of	prep.	就…而论
in the character of		以…的资格;扮演
in the event of	prep.	万一
in the face of	prep.	当…之前
in the hope of		希望
in the interests of		为…打算
in the light of		按照
in the long run	adv. phr.	终于,毕竟

in the main	adv. phr.	大体上,基本上
in the matter of	adv.	关于
in the presence of	adv.	在…之前,在…面前
in the rear of	adv.	在…之后
in the same sense	adv. phr.	在同样意义上
in the way of	adv. phr.	在…意义上
in this connection		在这方面
in token of		作为…的记号
in turn	adv. phr.	依次,又,转向
in use	adj. phr.	通行
in view of	adv. phr.	鉴于,由…看来
in virtue of	adv.	借助于…,依靠
in respective of	adv.	不管,不顾
keep watch	v. phr.	值班,看守
known as		称为
lack of	n. phr.	缺少
later on	adv. phr.	今后,过后
lead to	v. phr.	导向,导致
let alone	conj.	至于…就更不必说了
likened to	v. phr.	被比喻成
little by little	adv. phr.	渐渐
lose sight of	v. phr.	未看见,忽略了
made of		以…制成
make out	v. phr.	了解,证明,成就
make(set)the pace	v. phr.	调整步子,调整速力
make sense of	v. phr.	了解…的意义
make trial of	v. phr.	试用,实验
make use of	v. phr.	利用
make up	v. phr.	弥补
many a time	adj. phr.	多次,几度
matter of fact	v. phr.	事实
mix with	v. phr.	混;混合
more and more		愈来愈多
more or less	adv. phr.	大约,有几分

mount… on…	v. phr.	把…安装在…
neither more nor less than		不多于也不少于,和…完全一样
no end of	adj. phr.	无数的,无限的
no longer	adv. phr.	不再
no matter	adj. phr.	无关紧要
no matter how much	adv. phr.	不论多少
no matter when	adv. phr.	不论何时
none other than	adv. phr.	不外是
nothing but	pron. phr.	不过,只
not only… but also	conj.	不但…而且
not so much as		与其说是…,不如说是…,甚至于不
not the least		全无
now and again	adv. phr.	时常,时时
now and now	adv. phr.	时而
now and then	adv. phr.	时时
of course	adv. phr.	当然
on a large scale		大规模
on account of	prep.	由于;基于
on all accounts	adv. phr.	无论如何
on behalf of	adv. phr.	代表…
once again	adv. phr.	再一次
once more	adv. phr.	再一次
once (and) for all		限此一次,断然,爽爽快快
one after another	adv. phr.	一个接着一个
one and the same	n. phr	同一
one by one	adv. phr.	一个个的
on no account	adv. phr.	决不
on the average	adv. phr.	平均说来
on the basis of	adv. phr.	基于
on the condition that		以…为条件
on the contrary	adv. phr.	反之
on the face of it	adv. phr.	外观上,似乎
on the ground of	adv. phr.	以…为理由
on the instant	adv. phr.	立即

on the on hand	adv. phr.	在一方面
on the other hand	adv. phr.	在另一方面
on the part of	adv. phr.	在…方面
on the point of	prep.	将要
on the score of		以…之故;因为
on the strength of	prep.	靠着,凭
on (the) top of	prep.	在…的顶上
opposite to	adj. phr.	正相反的;不相容的
other than	prep.	不同于,而不是
out and out	adj. phr.	完全;十分
out of	prep.	起源;由于;引用
out of order		出毛病
outside of(=outside)		在…外边
over against		正对着
over and above		此外
over and again		加上
over and over	adv. phr.	再三
owing to	prep.	因为,由于
parallel to (with)		平行,并行的
part from		和…分手
pass by	v. phr.	经过;看漏
pass on	v. phr.	前进;传递
pay attention to	v. phr.	注意
per cent	n. phr.	按百分计算,百分之…
peculiar to	vt. phr.	限于…,…特有的
pertinent to	adj. phr.	关于
pick up	v. phr.	拾起,收听
play a role	v. phr.	起作用
preferable to	v. phr.	优于,胜过
prefer … to	conj.	宁要…不要…
prevent…from	v. phr.	阻止…使不…
previous to	prep. phr	以前的,前的
prior to	adv. phr.	在…前
protect…from	v. phr.	保护…;使不受…侵害

pursuant to		按照…,依据…
put in force		实行
put out	v. phr.	灭火;出产;长出
range from	v. phr.	分布在
rather than	adv. phr.	而不
refer to	v. phr.	关于,参考,指
regardless of	prep.	不管,不顾
relate to	v. phr.	关系;论及
relative to	prep.	相对于,相关的
remain with		属于,赖于
responsible for	adj. phr.	对…负责
result from	v. phr.	由…引起
result in	v. phr.	结果形成,归于
run out of	v. phr.	用完
set about	v. phr.	开始,着手
set against	v. phr.	比较,对照
set aside	v. phr.	取消
set forth	v. phr.	显示,说明
set up	v. phr.	建立,创立,竖起
short of	adj. phr.	缺少
side by side with	adv. phr.	同…并排,与…并列
similar to	adj. phr.	像…;类似
so far as	conj.	至于;关于,就…而论
so forth		等等
so long as		只要
so…that		如此…以致
spoken of as		被称为
suitable to		适合于
subjected to	v. phr.	受到,遭遇到
superior in		在…方面较多(优)
superior to	adj. phr.	胜过,优于
table of contents		目次
take after	v. phr.	相似
take for	v. phr.	认为

take in	v. phr.	包括
taken together		总计
take into account	v. phr.	计入
take no account of		不计及
take off	v. phr.	拿去;脱(帽子和衣服等);起飞
take part in	v. phr.	参与
take place	v. phr.	发生;举行
take the place of	v. phr.	代替
take up	v. phr.	拿起;占(地位);开始
tend to	v. phr.	倾向(于)
thanks to	prep.	幸亏;由于
the more … the more …		愈…愈…
to a large(great) extent	n. phr.	很大程度上,大大
to some extent	n. phr.	到某种程度
to such an extent that	n. phr.	甚至
to the extent of	n. phr.	到…程度
turn off	v. phr.	扭关(电门,龙头等)
turn on	v. phr.	扭开(电门,龙头等)
turn out	v. phr.	制造出
turn to	v. phr.	依赖;变成
up to	prep.	迄,直至,到;从事,做,胜任
up to date		直到如今,现代化
use up	v. phr.	用尽,用完
vary with	v. phr.	照…变化;跟着…起变化
with a view to	adv. phr.	以…为目的,意在
within reach of	prep. phr.	力所能及的,手所能达到的
without regard to	adv. phr.	不顾…
with reference to	adv. phr.	以…为基准;关于
with regard to	adv. phr.	关于…,提及
with respect to	adv. phr.	关于…,提及
with the exception of	prep.	除…外
with the intention of	prep.	以…为目的,打算
with the object of	prep.	以…为目的
with the purpose of	prep.	以…为目的
work for	v. phr.	争取…,为…而工作

5. 科技英语的翻译

一般认为翻译要做到"信、达、雅",即要做到译文忠实于原文,正确、完整地表达原作的内容,译者不得任意歪曲、增删、篡改、遗漏。同时要求译文通俗易懂、文理通顺。并且必须使用专业术语和习惯表达方式。

翻译是理解和表达的紧密结合和辨证统一。翻译时首先要弄通原文的技术内容,然后用汉语正确表达出来。一般来说,理解越深刻,表达就会越正确。

翻译可分粗译和精译两步。粗译是在理解的基础上,将原文译成初稿,然后钻研翻译过程中的"疑案"。解决了这些疑难问题后,便开始精译。也就是在了解全文内容的基础上,进一步深入推敲词义和正确的汉语表达方式,将译文作细致的调整。最后还必须将译文与原文作仔细的校对。

下面简要地说明翻译中的一些基本原则。

5.1 要准确选择词义

英语词汇常常一词多义,例如"round"可有巡视(名)、绕行(动)、圆形的(形)、循环地(副)、在…周围(介)等不同含义及词性。所以要根据该词在句中的词性、作用、上下文联系及词的搭配来确定词义。例如:

The round earth rounds the sun.

显然,前一 round 是圆形的意思,后一 round 则是绕行的意思。

In the westerly wind aloft which circles the globe in middle latitude …

其中的"circle"虽然也有多义,但此处显然是环绕的意思。

在选择词义时,还要考虑汉语习惯,例如:

The front is associated with a sharp line of wind change.

这里 sharp 可译成"明显的",而不用"陡峭的"。

5.2 可适当引申词义

英语与汉语表达方式不同,不能逐字硬译,否则就会使译文生硬晦涩,含糊不清,不能确切表达原意,甚至文理不通或使人误解。因此须根据上下文将词义适当引申。即根据该词所处的语言环境,从意义上、习惯上、逻辑上引申出能表达该词的内在含义的新词义。需要引申的可以是单词或词组,也可以是整个句子。例如:

Long-range weather forecasting begins after the sequence of ordinary day-by-day forecasts has lost its margin of accuracy above simple persistence.

直译:所谓"长期"天气预报是继普通逐日预报失去了高于简单持续性预报的准确

限度时接着开始的。

引申后的翻译:所谓"长期"天气预报是在普通的逐日预报不再具有(应有的)优于简单持续预报的起码精确度之后着手进行的。

5.3 可适当增减词语

在同一句子中,英、汉语的词量不可能完全相等,一般都需作适当增减。这不仅不损害原意,而且可使译文更确切、通顺。增词的原则是增加一些原文中无其形而有其义的词。例如上例译文中的"原有的"就属于这种情况。而对于英语中必须而在汉语中多余的词,(如某些冠词、连词等)则可省略。

5.4 可适当重复词语

有时为了强调词语或使语句生动、确切可适当重复词语。例如:

The extended forecasts are coming to depend more and more on the new methods of short-range forecasting by direct numerical solution of the equations governing atmospheric motion.

延伸预报变得越来越依靠那些通过直接数值求解支配大气运动的方程而作出短期预报的新方法。

Water is the same substance whether in solid, liquid or gaseous state.

无论水处于固态、液态或气态,它都是同一种物质。

5.5 可适当改变句子成分或转换词性

有时为了适应汉语习惯,可适当改变句子成分或转换词性。例如:

The values of curve S of Fig. 3, however, are not representative of the actual heat expended by the surface.

然而,图3中曲线S的值不代表地面实际的热量消耗(表语译成谓语)。

There are three main laws of mechanics, or three laws of Newton.

力学有三大定律,即牛顿三定律(定语译成主语)。

Ice is not so dense as water and therefore it floats.

冰的密度比水小,因此能浮于水面(表语译成主语)。

It is obvious that a typhoon has considerably more kinetic energy than a extratropical cyclone.

显然,台风的动能比温带气旋的大得多(宾语译成主语)。

There is a large amount of energy wasted due to friction.

由于摩擦而损耗了大量能量(定语译成谓语)。

The letter F is commonly used for force, m for mass, a for acceleration.
通常用 F 表示力，m 表示质量，a 表示加速度（状语译成谓语）。

This explanation is against the natural laws.
这种解释违反自然规律（表语译成谓语）。

The expression of the relation between force, mass and acceleration is as follows.
力、质量和加速度之间的关系表示如下（主语译成谓语）。

Numerical weather prediction is the prediction of weather phenomena by the numerical solutions of the equations governing the motion and changes of conditions of the atmosphere.
数值天气预报是一种通过求解支配大气条件变化和运动的方程组而作出的预报（名词译成动词）。

By power it is meant the speed, or rate, of doing work.
所谓功率，指的是做功的速率（主语译成宾语）。

The larger a system is in scale, the longer it is in life.
系统的尺度越大，生命也就越长（主语译成宾语）。

Liquids are different from solids in that liquids have no definite shape.
液体不同于固体，因为液体没有一定的形状（宾语从句译成状语从句）。

One must have studied hard before one could succeed in mastering a foreign language.
一个人必须勤学苦练，才能精通一门外语（从句译成主句）。

Start a sound into the air, and it will make waves which you can not see.
如果向空中发出一个声音，它就会产生你看不见的波（并列句译成复合句）。

These waves, which are commonly called radio waves travel with the velocity of light.
这些波以光速传播，它们通常被称为无线电波（复合句译成并列句）。

5.6 被动语态的译法

科技英语中被动语态用得很多。当着重指出动作的承受者，或不必说明谁是动作的执行者时，可用汉语被动语态翻译，若不是特别强调被动动作时，则一般译成汉语主动语态。例如：

Water can be changed from a liquid into a solid.
水能从液体变成固体。

A certain amount of energy is required to make this change.
这种改变需要耗费一定数量的能量。

有的被动语态可译成汉语中的无主句。如：
The pressure can be determined provided that the temperature and the density are known.
只要知道温度和密度就可知道气压。
有时为了强调，也可把英语被动句译成汉语被动句。如：
The term will be discussed in next section.
该项将在下节中加以讨论。

5.7 倒装句及后置装饰的译法

科技英语中常见倒装句，例如：
Atmospheric pressure decreases with increase in altitude and so does the density of the atmosphere.
这种倒装句在译成汉语时，常用自然顺序，例如上句可译成：大气压随高度增大而减小，大气密度也如此。
有时为了特别强调或使句子平衡，上下文紧密衔接，亦可按倒装词翻译。如：
To the platform are attached the thermograph and the anemograph.
可译成：装在观测平台上的是温度计和风速计。
英语中的分词短语、不定式短语、介词短语、形容词短语等在作定语时，通常位于其所修饰的名词之后（即"后置修饰"），在译成汉语时这些后置修饰大多成为前置修饰。例如：
The rate of motion of an object is given in units of length and time.
一个物体的运动速率是用长度和时间的单位表示的。
注意：作为定语的各短语先译最远的，其次再译倒数第二、第三…而构成一个定语并放在所修饰的名词之前。

5.8 定语从句的译法

定语从句有三种译法：
(1)将主句与限制性定语从句融合译出。例：
Air moves from places where the pressure is high to places where the pressure is low.
空气从压力高处向压力低处流动。
(2)将非限制性定语从句从主句中分开，译成并列句或状语从句。例如：
Inertia is that property of matter because of which force is needed to accelerate a body.

惯性是物质的一种属性,由于这一属性,如要使物质产生加速度,就必须施加一个力(译成并列句)。

One obvious example is the elevation angle of the sun, which produces the normal march of the seasons.

一个明显的例子是太阳高度角,它产生季节的正常推进(译成并列句)。

A gas occupies all of any container in which it placed.

气体不管装在什么容器里,都会把容器充满(译成让步状语从句)。

(3)当定语从句较简单时可译成简单句。例如:

There is no place on the earth where the days in winter are longer than in summer.

地球上没有一个地方的冬天的白昼比夏天长。

5.9 长难句的译法

如第三节所说,英语中的长难句子都是由基本句型扩展或变化而来的,翻译时必须抓住全句的中心内容,弄清句子的构成和语法关系。长难句一般有三种译法:

(1)顺译法。当原文叙述层次与汉语相同时,可顺序译出。例如:

These methods, and the electronic computers requried to use them have not yet been developed to the stage where they might produce forecasts of departures from normal of the average weather for a couple of weeks, a month or a season.

这些方法以及应用这些方法所要求的计算机都还没有发展到可以制作几周、一月或一季的气象距平预报的阶段。

(2)逆译法。当原文叙述层次与汉语相反时就须逆原文顺序而译。例如:

Weather predictions beyond a two-week period remain unreliable despite technological advances.

尽管技术进步了,两周以上的天气预报仍然是不可靠的。

(3)分译法。当长难句中的从句或分词短语与主句内容联系不很密切时,可把它们译成短句。例如:

In the meantime, as a result of his hemispheric observations of winds over the oceans, Maury(1855) proposed a new model of the meridional circulation which, in his opinion, could also account for the middle-latitude prevailing westerlies.

与此同时,莫里(1855)根据对半球海洋上风的观测提出了一个新的经向环流模式。按照他的说法,这个模式还能说明中纬度盛行西风带的成因。

5.10 数量增减及倍数的译法

英语中表示数量增减及倍数的句型很多。下面列举的是常用句型及其译法。

(1)(A)as high(much,long…)as N

译法:(A)高达(多达、长达…)N。

例:The temperature in summer is as high as 35℃。

夏季温度高达35℃。

(2)(A) is N (B)

译法:(A)是(B)的 N。

例:The volume of the moon is $\frac{1}{49}$ that of the earth.

月球体积是地球的$\frac{1}{49}$

(3)as many(B) as (A)

译法:有多少(A)就有多少(B),或(A)如(B)那样多。

例:Normally there are as many electrons as protons in the atom.

通常原子中有多少质子就有多少电子。或:通常,原子中质子的数量和电子的数量相等。

(4)(A) as many again as (B)

译法:(A)是(B)的两倍;或(A)比(B)多一倍。

例:Cyclone A moves as fast again as cyclone B.

气旋A移动得比气旋B快一倍。

(5)(A)again as many as(B)

译法:(A)是(B)的两倍;或(A)比(B)多一倍。

例:Coriolis parameter is again as many as $\omega\sin\varphi$.

柯里奥利参数等于$\omega\sin\varphi$的两倍。

(6)(A) half as many again as (B)

译法:(A)是(B)的一倍半,或(A)比(B)多半倍。

例:High level wind speed is half as fast again as the cloud mass.

高层风速比那个云团快半倍。

(7)(A) half again as many as (B)

译法:(A)是(B)的一倍半,或(A)比(B)多半倍。

例:The life of this storm is half again as long as that storm.

这个风暴的生命期比那个风暴长一半。

(8)(A) increase by N

译法:(A)增加(了)N。

例:The speed of new computer has been increased by two times as compared with that of the old one.

新的计算机速度比老的计算机提高了2倍。

(9)(A)by N more than(B)

译法:(A)比(B)大 N。

例:The antenna is by 1 meter higher than that one.

这根天线比那根要高1米。

(10)(A)N more than (B)

译法:(A)比(B)多 N。

例:The average annual precipitation amount of place A is twenty per cent more than that of place B.

A 地的平均年降水量比 B 地的要多 20%。

(11)(A) increase to N

译法:(A)增加到 N。

例:The automatic weather stations have increased to 50.

自动气象站增加到 50 个。

(12)N% increase of (A)

译法:(A)增加 N%。

例:There is a 20% increase of typhoon in this year as compared with last year.

今年的台风比去年增加 20%。

(13)(A) increase N as (B)。

译法:(A)比(B)增加了(N-1)倍;或(A)是(B)的 N 倍。

例:In summer the rainfall increases 10 times as against in winter.

夏季雨量比冬季增加了9倍。或:夏季雨是冬季的10倍。

(14)(A) increase by a factor of N

译法:(A)增加(N-1)倍。

例:The rainfall exceeds the annual average by a factor of 2.

该雨量超过年平均1倍。

(15)(A) more by a factor of N

译法:(A)多(N-1)倍。

例:The daily range of temperature at the surface of Tibetan Plateau is greater by a factor of 9 than that of the eastern China.

西藏高原地面温度的日较差要比中国东部地区大8倍。

(16)(A)N as great as (B)

译法:(A)比(B)大(N-1)倍;或(A)是(B)的 N 倍。

例：The daily range of temperature at the surface of Tibetan Plateau is 21 times as great as that of the free atmosphere at the same level with the plateau.

西藏高原地面气温日较差要比与高原相同高度上的自由大气中的气温日较差大 20 倍。

注意： 当 as fast (much, many heavy, long…) as 前不是倍数而是分数时，就不译成"快、多、重、长…"，而可能译成"慢、少、轻、短…"。例如：

Pattern A situation is four-tenth as frequent as Pattern B.

A 型形势比 B 型形势少十分之六。

另外，若倍数是一个相当大的近似值，差一倍没有多大意义，一般可以照译不必减一。例如：

The sun is 330,000 times as large as the earth.

太阳比地球大三十三万倍。（或太阳的大小是地球的三十三万倍）。

(17)(A) reduce by N

译法：(A)降低(了)N。

例：Due to the shelter-belt the sand haze has been reduced by 60%.

由于有了那条防风带，沙减少了百分之六十。

(18)(A) reduce to N

译法：(A)降低到 N。

例：The rate of nocturnal rain in Lasa Vally is as high as 80%, while it is reduced to 60%—70% in northern and eastern Tibet.

在拉萨河谷夜雨率高达 80%，而在藏北和藏东夜雨率降至 60%～70%。

(19)(A) reduce N

译法：减少了 $\frac{n-1}{n}$，或减少到 $\frac{1}{n}$。

例：The false alarm ratio of tornado of the new forecasting method is reduced 3 times.

新预报方法的龙卷虚假警报率减少了三分之二(或：新预报方法的龙卷虚假警报率减少到三分之一)。

(20)(A) N as small as (B)

译法：(A)比(B)小 $\frac{n-1}{n}$，或(A)是(B)的 $\frac{1}{n}$。

例：The electric field at 10 km above the surface is about ten times as weak as that at surface.

地面上空 10 千米的电场比地面电场大约要弱十分之九。（或地面上空 10 千米的电场是地面电场的十分之一）。

(21) reduce by a factor of N

译法：降低 $\dfrac{n-1}{n}$；或降低到 $\dfrac{1}{n}$。

例：The new numerical prediction method will reduce the error probability by a factor of 9.

新的数值预报方法将使误差概率降低九分之八。（或新的数值预报方法将使误差概率降到九分之一）。

The frictional drag reduces the horizontal velocity by a factor of e.

摩擦拖曳力使水平速度减少到 $\dfrac{1}{e}$。

(22)(A) N-fold reduction

译法：减少了 $\dfrac{n-1}{n}$，或减少到 $\dfrac{1}{n}$。

例：One of the effects to seed a storm is about two-fold reduction in total amount of lightning.

对一个风暴播云的效果之一是使总的闪电数减少约二分之一。

(23)(A) N less than (B)

译法：(A)是(B)的 $\dfrac{1}{n+1}$（N 代表倍数），或(A)比(B)少 N（N 代表整数）。

例：A is twice less than B.

A 是 B 的三分之一。

X is five less than Y.

X 比 Y 少 5。

(24)(A) half as many as (B)

译法：(A)比(B)少一半。

例：The storm produced half as much rainfall as that one.

这个风暴产生的降水比那个风暴产生的少一半。

(25)(A) twice thinner than (B)

译法：(A)比(B)少一半。

例：This kind of film is twice thinner than ordinary paper.

这种薄膜的厚度只有普通纸的一半。

5.11 近似值的译法

科技英语中，常用下列方法表示大概数：

(1)用 over, above, more than＋数字，或倍＋odd 表示"…多"、"超过…"、"…以上"

等意思。如：

over five days	五天以上
above 3,000 meters	3,000 米以上
more than 30 degree	30 多度
twenty odd	20 多

(2)用 under, below, less than＋数字表示"…以下"，"不足于…"，"不到…"，"少于…"等意思。如：

less than 35	35 以下
under 100	100 以内
below 10 ℃	低于 10℃

(3)用 some, about, toward(s), nearly, more or less＋数字(或数字＋or so)表示大约、上下、左右、将近、几乎等意思。如：

some 2,000 years after Aristotle	在亚里士多德之后大约 2,000 年
about 350 B. C.	公元前 350 年左右
toward(s) 10 o'clock	将近 10 点钟
nearly one third	接近三分之一
more or less 2 weeks	2 周左右
850 hPa or so	850 百帕左右

5.12　数词短语的译法

有些含有数词的短语往往不表示具体数量，可看成惯用法，并有固定译法。例如：

in twos and threes	三三两两
two by two	两个两个的
second to none	首屈一指
within a factor of ten	在一个数量级上
fifty－fifty	各占一半，平均
on second thought	重新考虑
a thousand and one	无数的，许多
a hundred and one	许多，无数的
last but one	倒数第二
last but two	倒数第三
by hundred percent	全部地
a few tenths of	十分之几，有几成
at sixes and sevens	乱七八糟

the second half	后一半
by ones or twos	三三两两
four figures	四位数
twenty to one	十有八九

5.13 不定数量的译法

(1)某些代词或词组可表示不定数量,如:

anything of …	很少;一点
a bit of …	少量;一点
a crowd of …	许多
a few …	一些;少量
a good few …	相当多
a good many …	很多
a great quantity of …	大量;许多
a great deal of …	很多;大量
a handful of…	少数
a large amount of …	大量;许多
a large number of	许多;大量
a little…	一些;不多
a lot of …	许多;大量
many (of) …	许多
much (of)…	许多
a multitude of…	许多;大量
a number of	许多;若干
a particle of	少量;一点
a portion of …	若干;一部分
plenty of …	许多;若干
a quantity of…	一些
a small amount of …	少量
a small quantity of …	少量
a store of…	大量
some few (little)	少量
a wealth of…	大量
a world of	许多;大量

(2)在 million, lot, thousand, number, hundred, score, ten 等词后加 s 或组成词组表示不定量。例如：

millions	千百万；数百万
lots of …	许多，大量
thousand upon thousands of…	无数，成千上万
numbers of…	若干，许多
many thousands of …	成千上万
scores of …	许多，如几十
hundreds of …	许多，几百
thousands of…	几千

5.14 量的尺度概念

科技英语中有很多词表示数量大小、频率高低以及概率大小。这些词所表示的程度（尺度）可用下表表示。

percentage guide	Quantity	Frequency	Probability Adverbs Adjectives	Probability Verbs
100% ↓ 0	all/every each most a majority (of) many/much a lot (of) enough some a number(of) several a minority (of) a few/a little few/little no/none any	always usual(ly) normal(ly) general(ly) regular(ly) often frequent(ly) sometimes occasional(ly) rare(ly) seldom hardly ever scarcely ever never	 certain(ly) definite(ly) undoubtedly probably(probable) likely perhaps possibly possible maybe unlikely	will is/are must/have to should ought to may might can could will is/are can +not could

6. 数学用语的译、读

科技英语中常用的数学符号和公式的译、读方法如下：

(1) ＋ plus 加、正的。

(2) － minus 减、负的。

(3) ± plus or minus 加或减，正的或负的。

(4) × multiplication sign 乘号。

(5) ÷ sign of division 除号。

(6) ＝ sign of equality 等号。

(7) $a=b$　　 a equals b　　 a 等于 b。

　　　　　　a is equal to b

　　　　　　a is b

(8) $a\equiv b$　　 a is congruous with b　　 a 全等于 b。

(9) $2+3=5$　　Two plus　　　　three　　equals　　　　five.

　　　　　　　Two and　　　　 three　　is/are　　　　five.

　　　　　　　Two add　　　　 three　　is equal to　　five.

　　　　　　　Two added to　　three　　makes　　　　five.

　　　　　　　Two increased by　three　gets/gives　　 five.

　　　　　　　Two increased by　three　the result is　 five.

(10) $10-3=7$　　Ten minus three equals seven.

　　　　　　　　decreased by leave

　　　　　　　　take away

　　　　　　　　loss

　　　　　　　　Three subtracted from ten

(11) $4\times 5=20$　　Four times five equals twenty.

　　　　　　　　　multiply

　　　　　　　　　multiplied by

　　　　　　　　　multiply 4 by 5 and the result is twenty

(12) $8/2=4$　　Eight divided by two equals four.

　　　　　　　Divide eight by two and you get four.

　　　　　　　When you divide eight by two you have four.

(13) $a\neq b$　　 a is not equal to b.　　 a 不等于 b。

　　　　　　　a is not b

(14) $a \approx b$　　a approximately equals b.　　a 近似于 b。

(15) $a \pm b$　　a plus or minus b.　　a 加或减 b。

(16) 77∶1　　the ratio of 77 to 1.　　77 比 1。

(17) 20∶5＝16∶4　　the ratio of 20 to 5　　20 比 5 等于 16 比 4。
　　　　　　　　　equals the ratio of 16 to 4
　　　　　　　　　twenty is to five as 16 is to 4

(18) $a > b$　　a is greater than b.　　a 大于 b。

(19) $a < b$　　a is less than b.　　a 小于 b。

(20) $a \geq b$　　a is greater than or equal to b.　　a 大于或等于 b。

(21) $x \rightarrow \infty$　　x approaches infinity.　　x 趋于无穷大。

(22) (　)　round brackets, parentheses 圆括号。

(23) [　]　square brackets 方括号。

(24) {　}　braces, curly brackets 大括号。

(25) ∴ therefore 所以。

(26) ∵ since, because 因为。

(27) $E = \dfrac{\dfrac{p}{a}}{\dfrac{e}{l}} = \dfrac{pl}{ae}$　　E is equal to the ratio of p divided by a to e divided by l,

is equal to the ratio of the product pl to the product ae. E 等于 p 除以 a 比 e 除以 l, 等于积 ae 比积 pl。

(28) x^2　① x square; x squared　　x 的平方(x 的二次方)。
　　　② x to the second power
　　　③ x raised to the second power
　　　④ the square of x
　　　⑤ the second power of x
　　　⑥ x to the second power

(29) $5^2 = 25$　① the second power of 5 is 25　　5 的平方等于 25。
　　　② 5 square is 25
　　　③ 5 to the second power equals 25
　　　④ 5 raised to the second power is equal to 25
　　　⑤ the square of 5 is 25

(30) y^3 ① y cube; y cubed y 的立方(y 的三次方)。
② y to the third power
③ y raised to the third power
④ the cube of y
⑤ the third power of y

(31) z^{-10} z to the minus tenth power. z 的负 10 次方。

(32) $\sqrt{4}=2$ ① the square root of 4 is (equals) 2   4 的平方根等于 2。
② the square root out of 4 is (equals) 2

(33) $\sqrt{a}$ the square root of a. a 的平方根。

(34) $\sqrt[3]{a}$ the cube root of a. a 的立方根。

(35) $\sqrt[5]{a^2}$ the fifth root of a square. a 的平方的 5 次根。

(36) $t=\sqrt{-1}$ the unit of imaginary numbers.   虚数单位。

(37) $n!$, the factorial of n. n 的阶乘。

(38) $\lg a$, $\log_{10} a$ common logarithm   常用对数。

(39) $\ln a$, $\log_e a$ natural logarithm   自然对数。

(40) (a, b) the open interval $a < x < b$ 开区间。

(41) $[a, b]$ the closed interval $a \leqslant x \leqslant b$ 闭区间。

(42) $(a, b]$ the interval $a < x \leqslant b$ 半开区间。

(43) $[a, b)$ the interval $a \leqslant x < b$ 半开区间。

(44) $\sum_{i=1}^{n}$, $\sum_{i}^{n}$ sum to n terms. n 项的和。

(45) $\sum_{i=1}^{n}$, $\prod_{i}^{n}$ product of n terms. n 项的乘积。

(46) $f(x)$ the function of x. x 的函数。

(47) $\lim_{x \to a} f(x)$ the limit of $f(x)$ as x approaches to a.   当 x 趋近于 a 时 $f(x)$ 的极限。

(48) Δy the increment of y. y 的增量。

(49) dy the differential of y. y 的微分。

(50) $\dot{z}$ first derivative of z. z 的一阶导数。

(51) $\ddot{z}$ second derivative of z. z 的二阶导数。

(52) $\frac{dz}{dx}$ first derivate of z with respect to x. z 对 x 的一阶导数。

(53) $\frac{d^2 y}{dx^2}$ second derivate of y with respect to x. y 对 x 的二阶导数。

(54) $W = F \times D$ W equals F multiplied by D, where W means work, F means force and D means distance. W 等于 F 乘以 D，此处 W 表示功，F 表示力，D 表示距离。

(55) $\frac{\partial f}{\partial x}, f_x$ the partial derivative of f with respect to x. f 对 x 的偏导数。注意，这里符号 ∂ 读作 round d 或 curly d。

(56) $\int f(x) \, \mathrm{d}x$ the integral of $f(x)$ with respect to x, the primitive of $f(x)$. $f(x)$ 对 x 的积分，$f(x)$ 的原函数。

(57) $\int_a^b f(x) \mathrm{d}x$ the definite integral of $f(x)$ from a to b. $f(x)$ 从 a 到 b 的定积分。

(58) I(z), Im(z) the imaginary part of z. z 的虚部。

(59) R(z), Re(z) the real part of z. z 的实部。

(60) $\bar{z}$ the conjugate of z. z 的共轭复数。

(61) $|z|$ the absolute value of z. z 的绝对值。

(62) sin   sine   正弦。

(63) cos   cosine   余弦。

(64) tan, tg tangent   正切。

(65) cot, ctg cotangent   余切。

(66) sec secant   正割。

(67) csc cosecant   余割。

(68) arcsin, $\sin^{-1}$ arc sine   反正弦。

(69) arccos, $\cos^{-1}$ arc cosine   反余弦。

(70) arctan, arctg, $\tan^{-1}$, tg^{-1} are tangent   反正切。

(71) arccot, arc ctg, $\cot^{-1}$, ctg^{-1} arc cotangent   反余切。

(72) arcsec, $\sec^{-1}$ arc secant   反正割。

(73) arccsc, $\csc^{-1}$ arc cosecant   反余割。

(74) sinh hyperbolic sine   双曲正弦。

(75) cosh hyperbolic cosine   双曲余弦。

(76) tanh hyperbolic tangent   双曲正切。

(77) coth hyperbolic cotangent   双曲余切。

(78) sech hyperbolic secant   双曲正割。

(79) csch hyperbolic cosecant   双曲余割。

(80) p'   p prime。

(81) 希腊字母的读法，如下表所示：

希腊字母		读法	
大写	小写	英　语	国际音标注音
A	α	alpha	[′aːlfə]或[′ælfə]
B	β	beta	[′beitə]或[′biːtə]
Γ	γ	gamma	[′gaːmə]或[′gæmə]
Δ	δ	delta	[deltə]
E	ε	epsilon	[′epsilən]
Z	ζ	zeta	[′zeitə]或[′ziːætə]
H	η	eta	[′eitə]或[′iːtə]
Θ	θ	theta	[′θiːtə]或[′θeitə]
I	ι	jota	[ai′outə]
K	κ	kappa	[′kæpə]
Λ	λ	lambda	[′læmdə]
M	μ	mu	[mjuː]
N	ν	nu	[njuː]
Ξ	ξ	xi	[ksai]
O	ο	omicron	[ou′maikrən]
Π	π	pi	[pai]或[piː]
P	ρ	rho	[rou]
Σ	σ	sigma	[′sigmə]
T	τ	tau	[tau]
Υ	υ	upsilon	[juːp′sailən]
Φ	φ	phi	[fai]或[fiː]
X	χ	chi	[kai]
Ψ	ψ	psi	[psiː]
Ω	ω	omega	[′oumigə]或[′oumegə]

7. 科技英语的写作

7.1 科技论文的组成部分及其写作要求

英语科技论文(或报告)通常包括标题、摘要、引言、正文、结论和建议、总结、致谢、参考文献及附录等组成部分。

1. 标题

标题要求简短明了,概括全篇,引人注目。一篇论文往往能设想好几个标题,要根据内容进行比较和选择。按惯例,英语标题第一个词和每个重要的词的第一个字母要大写。标题一般不超过12～16个词。并要求用词质朴、明确、实事求是,避免冗赘夸大,罕见难懂的字眼。当一个短标题不足以概括论文内容时,可加副标题,予以补充。

2. 摘要

联合国科教文组织制定的科技杂志准则规定,正式的科技论文,不论其使用什么语言,都必须附有一段英文摘要(abstract)。摘要放在论文前面。好的论文摘要应是全文的缩影,简短扼要(一般来说,一篇一般长度的论文摘要字数为125～250字左右),即摘要字数不超过全文字数的3%,并能独立使用,使那些对所论问题有所了解的读者,即使不看正文,也能一目了然,从而可以决定是否有必要详细阅读正文。

摘要一般要概括主要的研究内容、突出的研究成果及其意义。摘要虽然通常放在正文前面,但往往是在最后写作的。写摘要要使用正规英语和标准术语,避免写缩写字,一般用第三人称。摘要本身要完整。下面举几篇论文摘要作为例子:

例1:

The Computation of Equivalent Potential Temperature

DAVID BOLTON

Atmospheric Physics Group, Imperial College, London, England

(Manuscript received 1 August 1979, in final form 18 March 1980)

ABSTRACT

A simplified procedure is described for computation of equivalent potential temperature which remains valid in situations such as in the tropics where a term which is omitted in the derivation of the conventional formula can lead to an error of several degrees absolute. The procedure involves new empirical formulas which are introduced for the saturated vapor pressure of water, the lifting condensation level temperature and the equivalent potential temperature. Error are estimated for each of these, and results are compared with those obtained by the similar, but more complicated procedures of Betts and Dugan (1973) and Simpson (1978).

相当位温的计算

本文描述了一个计算相当位温的简化程序,它适用于热带一类的情况。在热带,常规公式推导中所省略的项可以导致绝对温度几度的误差。本程序包含计算饱和水汽压、抬升凝结高度温度以及相当位温的新的经验公式。对每一种计算都估计了误差。其结果与 Betts 和 Dugan(1973)以及 Simpson(1978)所得到的类似的但比较复杂的程序作了比较。

例 2:

Numerical Analysis of Atmospheric Soundings

JOHN D. STACKPOLE

National Meteorological Center, Weather Bureau, ESSA, Washington D. C.

(Manuscript received 30 December 1966, in revised form 8 February1967)

ABSTRACT

Numerical methods of calculation the pseudo-adiabatic characteristics of saturated air parcels are presented, based on the Rossby definition of the pseudo-equivalent potential temperature. With these methods it is then possible to perform routine automatic analysis of soundings. As examples, techniques for determining the lifted condensation level, the level of free convection, and the convective condensation level are presented.

大气探空的数值分析

本文根据 Rossby 假绝热位温的定义,提出了一种计算饱和气块假绝热特征量的数值方法。应用这种方法,就可能进行探空的日常自动分析。作为例子,给出了确定抬升凝结高度、自由对流高度以及对流凝结高度的方法。

例 3:

Mesoscale Air Motions Associated with a Tropical Squall Line

JOHN F. GAMACHE AND ROBERT A. HOUZE, JR.

Department of Atmospheric Sciences, University of Washington, Seattle 98195

(manuscript received 10 July 1981, in final form 7 December 1981)

ABSTRACT

Composites of radar and wind observations in a coordinate system attached to a moving tropical squall line confirm that such a squall system is composed of two separate circulation features: a convective squall-line region and a stratiform anvil region. The squall-line region is characterized by mesoscale boundary-layer convergence, which feeds deep convective updrafts, and mid-to-upper-level divergence associated with outflow from the cells. The anvil region is characterized by mid-level convergence, which feed both a mesoscale downdraft below the anvil and a mesoscale updraft within the anvil cloud. Before this study, the mesoscale updraft in the anvil cloud of the tropical squally system had been somewhat speculative, and both the anvil updraft and downdraft had been inferred only qualitatively. The occurrence of the anvil updraft is now proven and quantitative profiles of the mesoscale anvil updraft and downdraft have been obtained.

和热带飑线相联系的中尺度空气运动

在一个依附于移动的热带飑线的坐标系中,雷达和风的观测资料的合成分析表明,飑线系统是由对流性的飑线区和层状的砧区两部分组成的。飑线区是由中尺度边界层辐合——这种辐合供应深对流上升区——以及与单体外流相联系的中尺度辐散为特征的。砧区是以中层辐合为特征的。这种中层辐合既供应砧部下方的中尺度下沉气流,也供应在砧状云内的中尺度上升。在本研究之前,对热带飑线系统砧云内的中尺度上升已有一些推测,而对砧部的上升和下沉气流只有过定性的推论。现在砧上升气流已得到证实,并且已得到砧部上升气流和下沉气流的定量廓线。

例4:

An Objective Technique for Separating Macroscale and Mesoscale Features in Meteorological Data

ROBERT A. MADDOX

NOAA, Environmental Research Laboratories, Office of Weather Research and Modification, Boulder, CO 80303 and Department of Atmospheric Science, Colorado State University, Ft. Collions, CO 80523

(Manuscript received 23 March 1979, in final form 29 January 1980)

ABSTRACT

An objective technique for quantitative scale separation has been developed to study atmospheric circulations associated with large complexes of thunderstorms. The scheme utilizes two separate low-pass filter analyses of the same data set to extract mesoscale and macroscale signals. An objective analysis of the total meteorological field (with microscale variations suppressed) is recovered as the sum of the mesoscale and macroscale components. Case study examples demonstrate that the technique is indeed useful for studying mesoscale convective systems. It is shown that convectively forced mesoscale circulations may significantly perturb the environmental flow on scales large enough to be detected in synoptic upper-air data. The case studies also suggest that the analysis routines could be utilized in operational forecasting applications.

一种分离气象资料中大、中尺度特征的客观方法

本文提出了一种可用于研究与大的雷暴复合体相联系的大气环流的定量的尺度分离的客观方法。方案中应用了同一组资料的两个低通滤波分析以提取大尺度和中尺度信号。总气象场(隐含小尺度变化)的客观分析作为大、中尺度分量的总和而被重新获得。个例研究证实这种方法对研究中尺度对流系统是很有用的。对流强迫的中尺度环流可以明显地在大到足以在天气学高空资料中被觉察的尺度上扰动环境气流。个例分析也指出，这种分析程序也可以用于业务预报的应用。

以上举了几个摘要的例子，我们可以模仿这些写法，利用阅读材料中的课文来练习论文摘要的写作。

3. 引言

引言是科技论文的引子，它向读者说明论文的主题、目的、引起研究要求的现实情况、研究所涉及的界限、范围、研究历史、背景以及写作的计划、总纲和规划。下面两篇短文分别为上面已给出摘要的 David Bolton 和 R. A. Maddox 的论文的引言，可以作为我们学习引言写作的例子。

例1：

Introduction

The equivalent potential temperature (also known as pseudo-equivalent potential temperature) will be taken, as in Holton (1972), to be the final temperature θ_E which a parcel of air attains when it is lifted dry adiabatically to its lifting condensation level, then pseudo-wet adiabatically (with respect to water saturation) to a great height (dropping out condensed water as it is formed), then finally brought down dry adiabatically to 1000 hPa.

The first two processes are examined by Simpson (1978), the first by Betts and Miller (1975) and the second by Betts and Dugan (1973), hereafter referred to as SS, BM and BD, respectively. SS shows that a term normally neglected in the pseudo-adiabatic ascent can lead to errors of up to 3 K in the computed value of θ_E which would, for instance, result in a significant underestimate of the height to which penetrative convection can reach. Both SS and BD give approximate empirical formulas for correcting this error.

In this paper the formulas of SS and BD are examined and compared with simpler empirical formulas which are proposed, and which are found to permit a better fit to values computed by numerical integration. Estimates are made of the errors introduced at each of the stages for attaining the equivalent potential temperature, including the final dry adiabatic descent.

It should be noted that the definition of equivalent potential temperature given above is not the only possible one, since account could have been taken of heat retained by condensed water and of latent heat of ice formation, which can affect the value by several degrees, as shown in Saunders (1957). Inclusion of these effects, however, would require a detailed study of cloud microphysics, which is beyond the scope of this paper and, therefore, they have been omitted in keeping with common practice.

例2:

Introduction

The scales of midlatitude convective weather systems are usually defined to be on the order of diameter $\sim 1-10$ km for an individual thunderstorm; and length 100 km, width $10-100$ km for a squall line (Holton, 1972; La Seur, 1974). The average spacing of upper air observations over the United States is 400 km so that it seems valid to consider convective weather systems as subgrid phenomena in numerical analysis and prediction schemes. However, an ongoing study (Maddox, 1980) is documenting the frequent occurrence of convectively driven, mesoscale weather systems over the central United States whose dimensions exceed those defined above by more than an order of magnitude. Furthermore, Fritsch et al. (1979) and Maddox et al. (1979) have documented convective systems that appeared to modify their near environment on horizontal scales detected by synoptic upper air observations. One conclusion of this paper is that this type of large, meso-α scale $(250-2500$ km$)$, convective complex should not be treated as a subgrid feature.

It is important that we be able to evaluate the influences and controls of the large scale on the development and evolution of mesoscale convective complexes and, most importantly, the extent and magnitude of macroscale feedbacks and modificatins produced by these systems. An objective analysis technique which separates mesoscale meteorological features from the background macroscale environment has been developed. Its basis and example applications are presented in the following section.

4. 正文

正文是科技论文的主体。正文写作是将作者在科研中所形成的新鲜观点正确地表达出来的过程。文中要尽量压缩已为众所周知的议论,突出新的发现和观点。论述要直截了当。

5. 结论

结论是论文的总的观点,是实验和分析结果的逻辑发展,也是整篇论文的归宿。结论必须完整、正确、鲜明。结论不能用罗列成果代替,而必须是把分析结果推进一步。总之,结论要反映作者通过概念分析、判断、推理过程所形成的总观点。下面举的例子是上面提到的David Bolton文章的结论。

Conclusion

The following formula is recommended for computation of equivalent potential temperature for a water-saturation pseudo-adiabatic process:

$$\theta_e = T_K \left(\frac{1000}{p}\right)^{0.2854(1-0.28\times10^{-3}r)} \times \exp\left[(\frac{3.376}{T_L} - 0.00254) \times r(1+0.81\times10^{-3}r)\right] \quad (43)$$

where T_K, p and r are the absolute temperature, pressure and mixing ratio at the initial level, and T_L is the absolute temperature at the lifting condensation level, given by any of (15), (21) or (22). The maximum error in values thus obtained is 0.3 K, the main contribution to the error arising from neglect of variation of the specific heat of dry air with temperature and pressure, an error which also affects the value of the potential temperature θ.

6. 致谢

在研究工作中,若得到别人常规之外的帮助时,可在总结或在论文结束处表示感谢。注意用词要恰如其分。下面是两则致谢词可供参考。

例1:

Acknowledgments. I would like to thank Dr. A. K. Betts, Dr. K. J. Bignell, Mr. G. Dugdale and Mr. J. R. Prober-Jones for their helpful comments on this work.

例2:

Acknowledgments. Dr. E. J. Zipser of NCAR has assisted the authors on this study in many ways. Both authors have been visitors with Dr. Zipser's GATE group at NCAR during the course of this study. Mr. M. D. Albright and Mr. E. E. Recker provided the authors with rainsonde data in highly usable form. During the GATE Woods Hole Seminar in 1979, Prof. William Gray of Colorado State University strongly encouraged Prof. Houze to pursue the idea of composition winds with respect of the GATE radar data. This research has been supported by the Global Atmospheric Research Program, Division of Atmospheric Sciences, National Science Foundation and the GATE Project Office, National Oceanic and Atmospheric Administration, under Grant ATM78—16859. This paper is Contribution No. 590, Department of Atmospheric Sciences, University of Washington.

7. 参考文献

科技文献列举参考文献是传统惯例，反映作者严肃的科学态度和研究工作的广泛依据。凡引及其他作者的论文、观点、成果，都应在参考文献中注明。一般有两种注法。

第一种注法：在引证作者的姓名之后注上数码，然后参考文献栏中注明作者姓名、文献名称、期刊名，卷（期）数，或出版社名，出版年份及页码。

第二种注法：在引证的观点之后注上作者的姓名及作品发表的时间，然后在文献栏中作出和上述同样的详细注明。国外的论文大多采用这种注法。例如：在 David Bolton 的文章引言中提到：The first two processes are examined by Simpson (1978), the first by Betts and Miller (1975) and the second by Betts and Dugan (1973)…在文献栏中便必须给出上述三篇论文的详细注明，如其中最后一篇的注明如下：

REFERENCES

Betts, A. K. and F. J. Dugan. 1973. Empirical formula for saturation pseudo-adiabats and saturation equivalent potential temperature. J. Appl. Meteor. 12, 731—732.

7.2 有关写作的技术问题

1. 标点符号

（1）句号（.）

句号用在陈述句的结尾及表示缩写或省略。例如：Dr. , 8 a. m. , Ca. (California 的省略)。

（2）逗号（,）

①逗号用在连接一系列并列的词或片语之间。例如：

Condensation of water vapor takes place on tiny particles of salt, dust or smoke, called condensation nuclei, that exist in abundance in the air. (文献[7],p5)

②副词和副词片语要用逗号。例如：

To understand the general circulation of the air, however, one must also consider the rotation of the earth. (文献[7],p7)

③非限制性从句要逗号分开(限定性从句则不用逗号分开)。例如：

Except for radiation, which can be observed by weather satellite, the processes that produce and alter the weather cannot be measured directly but must be calculated from observations of atmospheric variables. (文献[7],p1)

④一个长的从句放在主句前,从句后要用逗号。例如：

If weather changes occurred according to a logical pattern, predictions could be made using mathematical calculations. (文献[7],p3)

(3)分号(;)

①一句中分几个主要部分,第一和第二两个主要部分之间已用了逗号,则用分号分隔以后的各个主要部分。例如：

Basically, the physical climatologist studies the exchanges and transformations of solar, atmospheric, and terrestrial radiant energy, and the storage of this energy as heat in soil, air, plants, animals and water; the exchanges and storage of water between and within these media in the form of rain, snow, evaporation, streamflow, soil water, and the water of plant and animal tissues; and the similar exchanges of carbon dioxide, oxygen, nitrogen and other rarer gases. (文献[7],p133)

②分号分隔并列句的组成部分,可省略连词 and。例如：

This work involves elaborate instrumentation, and is usually done at fixed sites where research equipment can be assembled; it is a form of applied experimental physics. (文献[7],p132)

(4)冒号(:)

①冒号用在一个长问句的前面。例如：

The crucial question then becames: will the evolutions of two nearly identical initial states of the atmosphere diverge in gross statistics as well as in detail, and if so, how rapidly? (文献[7],p70)

②冒号还用在一串词的前面。例如：

The energy that matter in physical climatolgy are: sensible heat, latent heat, kinetic energy, geopotential energy, electrical energy, etc. (文献[7],p134)

③冒号出现在比率和时间上。例如：

The ratio was 7:9 at 8:30 a.m.

(5)惊叹号(!)和问号(?)

科技论文中很少用惊叹号;问号的用法与普通英语相同。

(6)破折号(——)

①破折号用以表示思想突变。例如:

The Soviet emphasis was lofting larger and larger bodies——proof visible of the immense power of their rocket propulsion system. (文献[7],p294)

②破折号用以注解或说明。例如:

This long-wave emission is emitted by three gases——ozone, carbon dioxide, and mainly water vapor. (文献[7],p99)

(7)撇号(')

科技论文中,撇号用作所有格和符号、数字、字母的复数。例如:

four days' work; the z's; the x's…

(8)双引号(" ")

①双引号用作分隔主要的引语。例如:

Only a decade ago it was not uncommon to hear people say, "It should be nice tomorrow; the weather report calls for rain." (文献[7],p49)

②双引号起分隔新词的作用,并有汉语中的"所谓"的意思。例如:

①Variations of annual or even longer-term averages belong to the realm of "climatic fluctuation." (文献[7],p65)

②"Long-range" weather forecasting begins after the sequence of ordinary day-by-day forecasts has lost its margin of accuracy above simple persistence.

在以上两个例句中,climatic fluctuations 及 Long-range forecasting 都是新词,所以要用引号分隔。

(9)括号( )

一个完整的句子放在括号内时,句点放在后括号内,而当句子的一部分放在括号内时,句点放在后括号的外面。例如:

①The invention of the telegraph in 1840 spurred the establishment of networks of weather observing stations and national weather services in many countries during the later help of the century. (Today, international cooperation among these national weather services is effected through the WMO, a specialized agency of the United Nations.) (文献[7],p34)

②The core of highest wind speeds in the jet stream is found near the 2000-millibar level (about 40,000 feet or 12 km). (文献[7],p226)

(10)连字符(-)

在科技论文中连字符用得比普通英语多。连字符一般作连接复合字以及转行用。连接复合字的规律如下：

①两个（或两个以上）的词合成一个词。如：

first-order energy transfer

middle-latitude prevailing westerlies

small-scale motion

short-range information

high-speed computer

balloon-borne radiosonde

②包含一个前置词片语时。如：

up-to-the-minute data　最现代化的数据

out-of-use equipment　不能用的装备

③当第一个字为 self 时。如：

self-registering meteorograph　自记气象计

④区别于一个同形的字。如：

re-form(reform)

⑤没有连字符可使拼写或发音发生混乱。如：

skill-less　不是 skillless

2. 大写字母的用法

下面的情况都不用大写字母：

①名词与数目或字母在一起时。例如：

page 51；experiment A.

例外的情况如：

Table 3；Figure 4(或 Fig. 4).

②一年四季。

③引文的一部分，其第一字母不大写。例如：

The circulation theorems showed "the dynamical effects of density variations are important in the atmosphere and oceans."

下面的情况要大写：

①论文标题、表格名称及主要标题；

②名称（及其派生词）如 Spanish，Spain；

③第二层、第三层题名的第一个字母；

④普通名词用作名称，其简称的第一个字母要大写。例如：

Nanjing Institute of Meteorology 简称 Institute；

⑤冒号后的一句话、一个引语、一个字的第一个字母要大写。例如：

You may have asked:"Just how accurately is the weather currently being forecast?"

3. 数字和符号的写法

①一句话若以数字开头时,数字要用文字,而不用阿拉伯字码写出。例如：

19th-century observers learned that anticyclones are usually fair-weather areas, and that storminess, precipitation, and rapid temperature changes occur in cyclones. (文献[7],p2)

在上句中,开头的 19th-century 应写 Nineteenth-century。

②一句话不宜用符号开始。例如：

If $a=4.3$, c is infinite   不恰当

If $a=4.3$, then c is infinite   正确

③不要以等号"="作为一个句子的主要动词,等号可作从句的动词,例如：

When (7) is substituted in (8), $a=b$.   不恰当

When (7) is substituted in (8), one obtains $a=b$.   恰当

④数学公式里,表示同位时,不要加逗号。例如：

By using the equation, $a=b$.   不正确

By using the equation $a=b$.   正确

8. 科技英语的快速阅读

阅读通常是为了获取信息。为了有效地从英文文献中获取信息，必须提高阅读速度。对我国经过快速阅读初步训练的学生，一般要求 80～100 单词/分钟的速度。要提高阅读速度，须注意下列问题：

(1)要将词连成词组连贯地阅读，而不要逐词阅读。例如：
The World Meteorological Organization of the United Nations
联合国世界气象组织，虽由 8 字组成，但这是一个单一的概述，应立即在脑中反映这一概念。像这类词组很多，如：

Weather surveillance satellite	气象监测卫星
frame of reference	参考系
sea and land breezes	海陆风
mountain and valley winds	山谷风
geophysical hydrodynamics	地球物理流体动力学
the law of gravitation	万有引力定律
partial differential equation	偏微分方程

等等。平时就应将它们作为词组来记忆。

(2)要避免过多地依赖词典，遇到生词时可根据上下文所提供的线索以及词的构成来猜测和判断生词的词义。

(3)要避免不必要的语法分析，特别是一些简单的句子，更不必去作语法分析。

(4)要避免多余的翻译。翻译是在阅读理解的基础上用另一种文字将原文意思重新表达出来的过程。翻译的目的是为别人提供方便。所以翻译时要力求语意准确，用词贴切，行文流畅。为此就须字斟句酌，反复推敲。但如果阅读的目的，主要是自己获取信息，那么只要能正确理解原文内容。自己懂得就行了，不必用汉语表达。另外，我们还要逐渐训练不通过汉语翻译而直接用英语思维的能力。

(5)要避免声读。朗读和默读对于练习语音、语调以及加强记忆都很有用，但不利于提高阅读速度。所以在快速阅读时，只用眼、用脑而不要有一切发声器官(包括喉、舌、唇等)的有声或无声的活动。

(6)要扩大眼睛的视距。即增加两次停顿之间目光所见的词数。一般来说，视距愈长，阅读速度愈快，反之，则愈慢。若一次只看一个字，速度显然很慢。例如：

Some books are to be tasted, others are to be swallowed, and some few are to be chewed and digested.

这样看完这个句子要作 17 次停顿。若一次能看几个字(即一个意义相对完整的词

组),则停顿就会大大减少,速度就快多了。例如:同样上面这句话,我们可以这样读:

Some books are to be tasted others are to be swallowed, and some few to be chewed and digested.

一般来说,合适的视距正好是一个意义相对完整的词组。所以要充分发挥视距的作用,应学会熟练地辨认词组,按词组进行阅读,要使目光对词组形成某种条件反射。

(7)要抓住段落的主要思想,从而理解全文的中心思想。具体做法是:首先,辨认主题句(即一段文字中能表达其主要思想的句子),并通过辨认主题句来确定段落的主要思想。主题句可能出现在开段、收段或段落中部。一般要注意开段句及收段句。其次,在没有主题句时,通过浏览全文,来酝酿概括段中要点。

(8)要利用提示手段,理解段落之间和句子之间的逻辑联系。能起揭示作用的词和词组,有连接词(尤其是连接分句的连接词)以及某些副词和介词词组。某些句型也能起提示作用。能起提示作用的词和词组相当于一种特定的信号,预先告诉读者下文的去向和性质,甚至可由此推断下文将要陈述的基本内容。下表列出一些常见的提示性词和词组以及它们所提示的内容。

	因		果	
提示因果关系	because (of)	因为	hence	因此,从而
	for	因为	thus	因此,于是
	since	既然,由于	so	所以
	as	因为	therefore	所以
	now that	既然	consequently	因而
	seeing that	既然	as a consequence	因而,结果(是)
	considering that	鉴于,考虑到	as a result	结果(是)
	that is why	这就是…的缘故	accordingly	所以
			for that (this) reason	为此,因此
提示对比关系	but	但是	on the contrary	相反地
	however	然而	on the other hand	另一方面,相反
	nevertheless	可是,仍然	although (though)	虽然
	yet	但是,然而	even though	即使
	whereas	而	even if	即使
	still	但是,然而	in spite of	尽管
	otherwise	否则,不然	despite	尽管
提示系列或时间顺序	one	其一	then	其次,然后
	two	其二	next	下一步,紧接着
	three	其三	finally	最后
	first(ly)	第一	at last	最后,终于
	second(ly)	第二	lastly	最后
	third(ly)	第三		

	因		果	
提示总括关系	in general generally speaking in short in a word in brief	概言之,一般说来 一般说来 简而言之 总而言之 简而言之	to be brief briefly in all in sum in simple words	简而言之 简而言之 总之 总之,简而言之 简而言之
提示强调作用	notice that note that remember that	注意 注意 要记住	don't overlook it is important to know be sure to know	不可忽视 重要的在于了解 务必懂得

(9)有时可略读原文。略读是快读的一种。它与一般的快读不同之处是它只是有选择地阅读部分文字,而不是每句必读。只浏览全文,了解大意,而不追求细节。一般来说,通过略读,如能理解 50% 的内容就算达到了目的。略读的方法有以下几种:

①阅读每段的开段句;

②阅读文章的开始段及结束段(它们通常包含文章的主要思想或内容梗概);

③若主题句或主要段位于文章中部,便要很快从上往下看;

④若文章有内容提要,则只看提要和各段的主题句,便可了解文章的大意。

下面是几篇短文,供大家作快读练习。首先大家可以测试自己的阅读速度。然后,再测试对文章的理解程度。看看能否回答文章后面所提的问题。最后,再对照上述各点,来改进阅读方法,提高阅读速度和加深对课文的理解。

练习 1:

THE WEATHER

People are most interested in the weather in countries where the climate is varied and uncertain. Human lives often depend on weather conditions. But how much do you know about the causes of different kinds of weather?

1. What makes the wind blow?

The pressure of the atmosphere varies all the time. Air increases in volume as the temperature rises, and so a cubic metre of cold air is heavier than the same volume of warm air. As warm air is lighter, it rises. It is displaced at lower levels by colder air which moves in. Wind is simply the movement of air between high and low pressure areas. The bigger the difference between the pressures, the stronger the wind is. Atmospheric pressure is measured with a barometer.

2. How are clouds formed?

The moisture in the atmosphere is produced by the evaporation of water and by the breathing of living things. As water vapour is lighter than air, it rises. It goes on rising until it condenses. Then it can be seen in the form of clouds. The process of condensation continues until the water becomes too heavy to stay in the air and it falls as rain.

3. How is snow formed?

Currents of very cold air make the water vapour in clouds freeze. The clouds then consist of minute particles of ice. But these ice particles remain lighter than air until the temperature falls below a certain point. Then they combine, become heavier, and fall as snow. Snowflakes are crystals with a beautiful, patterned structure.

4. What is lightning?

Lightning is a sudden discharge of electricity from cloud to cloud or from cloud to earth. The same sort of effect can be produced by connecting the positive and negative terminals of a storage battery. This creates a short circuit and a violent spark is discharged. Lightning from the sky is produced in the same way, usually between two clouds with opposite electrical charges.

Total number 327
Reading time _____
Words per minute _____
Comprehension score _____

(1) How many causes of different kinds of weather does the author mention in this passage?

(2) According to the author, why are people interested in the weather?

练习 2:

WHAT MAKES IT RAIN?

More nonsense is talked about the weather than about any other topic. This story, then, maybe nonsense. Or is it?

It is a story about Dr. E. G. Bowen, a scientist in Australia. While studying rainfall, Dr. Bowen noticed something curious that led him to an entirely new idea.

He noticed that rainfall records from 1902 to 1944 showed heavy rain on the same dates every six to eight years. For instance, in January there was a downpour on the

12th or 13th, the 22nd or 23rd, and around the 31st. It seemed strange that the dates were almost the same.

To make sure that it was not just chance, Dr. Bowen checked the Australian records back through another 50 years. The peaks of rain occurred on the same dates.

Was this true for the parts of the world? Dr. Bowen studied fifty-year records from South Africa, Chile, England and the United States. One by one, the records showed heavy rainfall on the same dates.

What scientific explanation could there be for this regularity?

Dr. Bowen figured that these effects could not be caused by anything here on the earth. They must be caused by something outside of the earth——something that happens on the same days of the year. Therefore, it must be somehow connected with earth's orbit around the sun.

Dr. Bowen could think of only one possibility: meteor showers. He listed the dates when the earth regularly passes through regions of space where there is meteor dust. He compared these dates with his storm charts. The dates of passage through meteor dust fell just 30 days before the peaks in rainfall!

Meteor dust is made of the remains of comets. The meteor particles range from the size of markables down to the finest dust.

Year in and year out, meteor dust is found in the same locations in space. Clouds of it do not wander from place to place. The earth passes through the dust on the same days each year. Astronomers know these dates. They are the dates for meteor showers.

So, Dr. Bowen suggested an interesting theory. We still do not know if his theory is correct. But no one has yet proved that the theory is not correct.

Meteor dust moves into the earth's atmosphere, some 50 or 60 miles above the surface. The bigger particles become white-hot streaks of light, or "shooting stars". The finer particles do not become "shooting stars". They sift through the air down toward the earth.

At many places over the surface of the earth there are no clouds, or clouds that hold only a little water. At such places the fine particles sift to the earth with no effect. But in other places the clouds are full of water droplets. When the meteor dust enters theses clouds, Dr. Bowen suggested, the droplets cluster about the dust particles. When the droplets grow large enough, they fall to the earth as rain or snow.

Of course this does not mean you can look at the date for meteor showers and predict that it will certainly rain. If Dr. Bowen is correct, your part of the earth can expect rain (or snow) on certain dates if cloud conditions are just right.

Total number 531
Reading time _____
Words per minute _____
Comprehension score _____

Reading Comprehension:

Circle the letter next to the best answer.
1. Heavy rain appears:
   a. very often.
   b. regularly on or close to certain dates in the year.
   c. seldom.
2. Dr. Bowen studied rainfall records:
   a. from Australia only.
   b. from the United States.
   c. from different countries.
3. This article indicates that:
   a. meteor showers have a great deal to do with comets.
   b. meteor showers are not comets.
   c. comets exist in the universe.
4. The author believes that meteor showers:
   a. occur at regular intervals.
   b. occur irregularly.
   c. are unknown.
5. It has been found that meteor dust is located:
   a. in different places in space.
   b. in the same places in space.
   c. close to the earth.
6. According to the information in this article:
   a. the author disagrees with Dr. Bowen.
   b. the author doesn't think Dr. Bowen's theory is correct.
   c. the author seems to be interested in Dr. Bowen's theory.

7. If the water droplets in a cloud become too large:
   a. they fall as rain or snow.
   b. they evaporate quickly in the atmosphere.
   c. they will freeze.

Word Study:

Fill in the blank in each sentence with the appropriate form of the word in bold type.

1. possibility: Life on Earth would not be _____ without rain.
2. prediction: People have tried to use different ways to a accurately _____ when and where it would rain or snow or be dry and windy.
3. regular: In the United States the National Weather Service _____ gives out weather information.
4. prove: Do the predictions _____ to be right?
5. cause: Cars are one of the major _____ of pollution.
6. finding: The fossil remain of tropical plants are _____ in Antarctica.
7. interesting: Young children are _____ in mystery stories.
8. correct: I always try _____ my own mistakes.
9. connection: There are many ways in which our ears behave as though they were _____.
10. strange: He looked at me _____.

练习 3:

AIR POLLUTION

Ever since early man lit his first fire he has been polluting or fouling, the atmosphere. It is however, only since the Industrial Revolution, in the last century, that pollution of the air we breathe has reached dangerous levels.

Air pollution arises form modern man's demands for energy——to light and heat his home, to run factories and to power vehicles and aircraft for travel.

Whenever a fuel like petroleum, oil, natural gas or coal is burned, it produces heat, which can be turned into power. But it also produces dirt and dangerous chemicals.

Burning these fuels produces many unwanted substances, such as smoke, and the gases sulphur dioxide and carbon monoxide. Complicated chemicals called hydrocarbons ——

some of which can cause cancer —— and acids and poisonous compounds are produced. Into the air in the United States are pumped every year 65 million tons of carbon monoxide, 23 million tons of sulphur compounds, 15 million tons of sooty and oily compounds, 12 million tons of dust, and 8 million tons of nitrogen compounds —— and these figures could very well double by the end of the century.

Smoke, sulphur dioxide and carbon monoxide are main pollutants of the air. Smoke is made of very tiny particles of solid tarry material, which float in the air. Under special weather conditions the particles may mix with water vapour in the air and cause fog.

Fog is the biggest air pollution killer: in the famous London smog of 1952 about 4,000 people died as a result of breathing the sooty fumes. The sooty particles stick in the lungs and cause severe coughing. For a person whose lungs are already strained because of disease, the fog can be fatal. Decompose partially, producing a haze that stings the eyes and makes breathing difficult. One of the hydrocarbons, called benzpyrene, has caused cancer in experiments with animals.

Lead is a known poison too, and although the levels in the air from automobile exhausts are still small, many countries are bringing in laws to control them.

Total number 585
Reading time _____
Words per minute _____
Comprehension score _____

Reading Comprehension:

Circle the letter next to the best answer.
1. There are three main pollutants of the air:
a. sulphur, hydrocarbon and nitrogen.
b. smoke, sulphur dioxide and carbon monoxide.
c. carbon monoxide, lead and hydrocarbon.
2. In this article the author says that:
a. fog may result in a lung disease.
b. fog may result in a skin disease.
c. heart diseases are more dangerous than lung disease.
3. According to the information of this article:
a. bad weather conditions contribute to fog.

b. water vapour causes fog.

c. automobile exhausts cause fog.

4. The polluted air in major cities is directly related to:

a. the weather.

b. man's travel.

c. the development of industry.

5. _____ are mainly responsible for carbon monoxide in the air according to this article.

a. Chemicals

b. Factories

c. Automobiles

6. In this article it is implied that:

a. air pollution is one of the most serious problems facing the United States.

b. the air in the United States contains less poisonous gases now.

c. nitrogen compounds could double by the end of the twentieth century.

7. Some paragraphs of this article show:

a. cigarette smoking is dangerous.

b. automobiles endanger health.

c. automobiles are controlled.

Word Study:

Fill in the blank in each sentence with the appropriate form of the word in bold type.

1. pollute: Sulphur dioxide is one of the main _____ of the air.

2. breathe: He had been running so hard that he was out of _____.

3. danger: Don't _____ the lives of the children.

4. directly: Please give me a _____ answer.

5. fatal: Three people were _____ injured, and two were seriously injured in the two-car collision.

6. occurrence: A total eclipse of the moon _____ two years from now.

7. die: The old man's _____ was due to natural causes.

8. personal: Who was that _____ you talked with?

9. high: What is the _____ of this building?

10. production: Many _____ are made from nylon.

练习 4：

FRONT

A front is a sloping surface of discontinuity in the troposphere, separating air masses of different density or temperature. The passage of a front at a fixed location is marked by sudden changes in temperature and wind and also by rapid variations in other weather elements, such as moisture and sky condition.

Although the front is ideally regarded as a discontinuity in temperature, in practice the temperature change from warm to cold air masses occurs over a zone of finite depth, called a transition of frontal zone. In typical cases the zone is about 3000 ft (1 km) in depth and 100mi (100-200 km) in width, with a slope of approximately 1/100. The cold air lies beneath the warm in the form of shallow wedge. Temperature contrasts are generally strongest at or near the Earth's surface. In the middle and upper troposphere, frontal structure tends to be diffuse, though sharp, narrow fronts of limited extent are common in the vicinity of strong jet streams. Upper-level frontogenesis is often accompanied by a folding of the tropopause and the incorporation of stratospheric air into the upper portion of the frontal zone.

The surface separating the frontal zone from the adjacent warm air mass is referred to as the frontal surface, and it is the line of intersection of the surface with a second surface, usually horizontal or vertical, that strictly speaking constitutes the front. According to this more precise definition, the front represents a discontinuity in temperature gradient rather than in temperature itself. The boundary on the cold air side is often ill-defined, especially near the Earth's surface, and for this reason is not represented in routine analysis of weather maps. In typical cases about one third of the temperature difference between the Equator and the Pole is contained within the narrow frontal zone, the remainder being distributed within the warm and cold air masses on either side.

练习 5：

CYCLONES AND ANTICYCLONES

The cyclone is an area of low pressure enclosed by roughly circular isobars, that is, where pressure decreases from its outer rim to its center. Such a system has long been known in meteorology as a cyclone.

Individual cyclones differ greatly in size, ranging in diameter from 100 to 2,000 miles, the average diameter being 1,000 miles or more. The typical round or elliptical low has, near the surface of the earth, moderate winds directed inward and around the center of low pressure, making angles of from 20° to 40° with the isobars. The direction or movement of air is counterclockwise in the Northern Hemisphere, responding to the influence of the earth's rotation. Such a movement of air around a center of low pressure is called cyclonic circulation.

The other characteristic pattern of isobars to be observed on almost any weather map is the high-pressure area or anticyclone. Isobars enclose an anticyclone in circular or elliptical fashion. Winds spiral outward around the center, clockwise in the Northern Hemisphere, gradually crossing the isobars toward low pressure. This system of diverging winds forms an anticyclonic circulation. Unlike a low-pressure center, which may be composed of two or more air masses, an anticyclone usually consists of a single air mass.

Answer the Following Questions:

1. How is a front defined?
2. What happens when a front passes a fixed location?
3. What is the usual depth of a frontal zone?
4. What is a frontal surface?
5. How is the temperature difference between the Equator and the Pole distributed?
6. What is a cyclone?
7. How is pressure distributed in a cyclone?
8. What are isobars?
9. Are the winds diverging or converging around a cyclone?

9. 学术报告和讨论

学术会议是科技交流的重要方式之一。这里介绍一些关于会议主持人、报告人以及与会的听众和讨论者的常用句型。

会议主持人宣布会议开始、介绍报告人及报告题目、欢迎重要来宾可用下列用语：

Ladies and gentlemen, I declare the meeting (conference) open.

Shall we getting started? Shall we get down to business?

I have great pleasure in introducing Mr. Li, our first speaker. Mr. Li will speak on …

It's a great honour for us to have Professor K with us in our group today.

报告人的发言要很好地组织。要有开场白、结束语；要有明显的主题、鲜明的层次；段落之间要有很自然的过渡；讨论过程中要正确地表述论点、论据、概述、定义、例子、图表、问题的焦点；要就正反面阐述进行对比、对照；要用另一种说法进一步说清问题；要加强论证；最后要作出结论和总结。

下面介绍是常用句型：

9.1 开场白（点明主题）

I'd like to talk about …

We're going to look at …

What I intend to consider $\begin{Bmatrix} \text{today} \\ \text{in particular} \end{Bmatrix}$ is …

We hope to cover …

I want to deal with …

9.2 段落过渡

Let's begin by looking at …

Well, let's continue by considering …

O. K, now let's consider …

So, move on to …

Now, I'd like to turn now to …

Right, pass on to …

Passing on to …

9.3 结束语

Well, I think we should stop here.
Well, I think that just about covers that topic …
I hope that gives you some idea about …
I hope that gives an outline of …
That's all I've got to say about … for today.
If you've nothing further to add, I think we should stop there.

9.4 指要点

The main $\begin{Bmatrix} \text{problem} \\ \text{conflict} \\ \text{question} \end{Bmatrix}$ is …

The most important point is that …

One of the main issues $\begin{Bmatrix} \text{is} \\ \text{is that} \\ \text{is whether} \end{Bmatrix}$ …

9.5 比较和对照

Unlike X, Y is …
In contrast to …
X differs from Y in that it …
X is …, Y, on other hand is …
Like X, Y is …
X is similar to Y in that it …
X is …, similarly, Y is …

9.6 以另一种方式重新阐述

Yes, in other words …
Stated another way, …
Well, as I see it, …
Well, to put it another way …
That is …
To put it another words …

In short …
That is to say …

9.7　增加理由、加强论证

Besides, …
What's more, …
In any case, …
Not only X, but also Y …
Not only X, but Y … as well.
And anyway …
Furthermore, …
Moreover, …
In addition, …

9.8　举例子

Take X for example, …
Look at X for instance, …
Consider example X.
One instance of this is …
Another case of this is …
You can see this sort of thing (situation) with X.

You can see(find) $\begin{cases} \text{an example of this} \\ \text{a case of this} \end{cases}$ in Y (place)

9.9　指示图、表，并通过图、表谈论问题

$\left. \begin{array}{l} \text{According to} \\ \text{As you can see from the} \\ \text{If you look at} \end{array} \right\}$ $\begin{Bmatrix} \text{information} \\ \text{data} \\ \text{figures} \\ \text{statistics} \end{Bmatrix}$ shown in this $\begin{Bmatrix} \text{diagram, …} \\ \text{chart, …} \\ \text{graph, …} \\ \text{table, …} \\ \text{histogram, …} \\ \text{map, …} \end{Bmatrix}$

As I have tried to show, $\begin{cases} \text{range in X is …} \\ \text{the increase in X is …} \\ \text{proportion of X is …} \end{cases}$

As this table shows, there is a $\begin{cases} \text{sharp fall in} \cdots \\ \text{steady increase in} \cdots \\ \text{even distribution of} \cdots \end{cases}$

The chart $\begin{cases} \text{suggests} \\ \text{indicates} \end{cases}$ $\begin{cases} \text{wide range of} \cdots \\ \text{steep rise in} \cdots \end{cases}$

The $\begin{cases} \text{left-hand column represents} \cdots \\ \text{bottom row indicates} \cdots \\ \text{broken line represents} \cdots \\ \text{vertical axis tells us} \cdots \\ \text{top part shows} \cdots \end{cases}$

9.10　表示层次

first
next, then
after that
meanwhile
in the meantime
at this stage
afterward
subsequently
eventually
finally
at the end

initially
after this
at this stage
following this
then
during this stage later
during this process during the process
at same time
subsequently
after this
at this stage
finally

9.11　下定义

例：A = a front; B = a sharp line of rapid wind and temperature changes.

A is　　　　　　　　　　　　　　B
A is defined as　　　　　　　　　B
A may (can, could) be defined as　B
B is known as　　　　　　　　　　A

B is labeled as   A
B is called (named, termed)   A
B may (could, can) be called (referred to) as A

9.12 做总结

To sum up, I think we …
In short, I agree that …
In all,
Altogether, } we thought that …
Overall,
We all agree that …
There is general agreement about …
There is still some conflict about …
Are we all agreed that …
What we're all saying is …
The main points that have been made are …

9.13 讨论

听众及讨论者的常用句型如下：
(1)要求了解细节或更多情况：

Could you { be more specific about …?
explain what this means in terms of …?
give us some more details about …?
explain what the advantages (disadvantages) of … are?
explain what the implications of … are?

(2)要求澄清或进一步解释问题：

I'm sorry but
Excuse me but } I don't quite { get   } { the point.
I'm afraid   follow   what you mean.
                                    understand what you're getting at.

I don't know (much) about that
I'm not quite sure   } { the point.
I don't quite see   what you mean.
                                       what you're getting at.

I wonder if you could   } { give some more details?
Do you think you could   go over it again?

Could you explain it a bit more?
What exactly do you …
Could you explain what you …
I don't understand what you mean by X.
I don't quite grasp what you …
I don't quite follow what you …
By X do you mean …?
Am I right in thinking that X is …?
Am I correct in assuming that X is …?
Don't you $\begin{Bmatrix} \text{think} \\ \text{agree} \end{Bmatrix}$ that …?

(3) 发表评论，陈述观点，表示同意或不同意：
In my opinion, …
As far as I can see, …
As I see it, …
I think that, …
It seems to me that …
I agree with you.
That's a good point.
I think you're right.
I feel the same.
$\left.\begin{matrix} \text{Exactly,} \\ \text{Yes,} \end{matrix}\right\}$ that's true (quite true).
Well, that's true.
Yes, $\begin{cases} \text{I take your point.} \\ \text{I see what you mean.} \end{cases}$
I'm afraid I don't agree.
I'm sorry but I can't agree.
I'm not sure about that.

回答者可用下列句型：
对于难回答的问题，需思考时间时
Let me think …
Well, …
Um, …

若只能作不肯定的回答时

As far as I know, …

Well, I'm not sure, …

Well, I don't know much about it, but I $\begin{cases} \text{think} \cdots \\ \text{believe} \cdots \end{cases}$

想要避免回答时

$\left.\begin{matrix}\text{I'm sorry but}\\ \text{I'm afraid}\end{matrix}\right\}$ $\begin{cases}\text{I don't know much about X.}\\ \text{I'm not an expert on Y.}\\ \text{that's not my field.}\\ \text{I've no idea.}\\ \text{I don't know.}\\ \text{I'm not sure.}\\ \text{I can't say.}\end{cases}$

10. 记笔记及缩略语

10.1 怎样记笔记

听报告或听课时记笔记有助于集中注意力听讲,并可供事后复习参考。阅读时作笔记有助于抓记文章要点,便于记忆。记笔记要快而简明,这主要通过减少句量、缩短句子和词来达到。具体做法是:

(1)省略例子和与主题无实质性关系的句子;
(2)把注意力集中在重要句子上;
(3)写短语而不必写全句;
(4)应用通用符号或记号及缩略语。

常用的符号及记号如下:

∴　　therefore
∵　　because
√　　correct
×　　wrong
?　　question (is the statement correct or not ?)
&. or +　　and, plus
¨ ¨　　ditto (means the same as the words immediately above the ditto marks)
＝　　is, are, have, has, equals
≠　　does not equal, differs from, is the opposite of
→　　leads to results in causes
↛　　does not lead to result in cause

常用缩略语

e. g.	for example
i. e.	that is
etc.	etcetera, and so on
a. m.	ante meridiem
p. m.	post meridiem
Sc.（Sci.）	Science
of.	compare
viz.	namely（即,也就是）
c.（or ca.）	about approximately

N. E.	note
C_{19}	nineteenth century
C_{20}	twentieth century
1920s	1920—1929
1980s	1980—1989
1st	first
2nd	second
3rd	third
Eng.	English
Q.	question
A.	Answer
df.	definition
approx.	approximately
dept.	department
diff.	(s) difficult(y) (-ies)
excl.	excluding (排除,排斥)
Fig.	Figure
govt.	government
imp.	important, importance
EC	Executive Committee (执行委员会)
eq.	equation
incl.	including
info.	information
lang.	language
ltd.	limited
max.	maximum
min.	minimum
no.	number
p. /pp.	page/pages.
poss.	possible/possibly
probs.	problems
re.	with reference to/concerning
ref.	reference
sts.	students

tho'	though	
thro'	through	
V.	very	
Sum.	summary	

10.2 气象科学中常用的缩略语词汇

A	absolute temperature	绝对温度
A	acceleration	加速度
Ac	alto cumulus	高积云
APE	available potential energy	有效位能
APT	automatic picture transmitter	图像自动发送器
A. S. L.	above sea-level	海拔高度
ASP	atmospheric sounding profile	大气探测廓线
atmos.	atmospheric(al)	大气的
avg(e)	average	平均的
AWRS	airborne weather radar system	机载气象雷达系统
AWS	American Weather Service	美国气象局
BL	boundary layer	边界层
BPF	band pass filter	带通滤波器
BUOY	buoyancy	浮力
℃	degree of Celsius	摄氏度
C	centigrade	百分度,百分温标
cal.	calorie	卡,卡路里
CB	cloud base	云底
Cb	cumulonimbus	积雨云
CBL	convective boundary layer	对流边界层
CC	cirrocumulus	卷积云
CCL	convective condensation level	对流凝结高度
CF	Coriolis force	柯氏力
C. G. S.	centimetre-gram-second	厘米·克·秒制
Ci	cirrus	卷云
CI	convective instability	对流不稳定性
CISK	conditional instability of the second kind	第二类条件不稳定
CMB	Central Meteorological Bureau	中央气象局
CMS	Chinese Meteorological Society	中国气象学会
coef.	coefficient	系数
conv.	convergence	收敛,辐合

conv.	conveyor	传输带
Cs	cirrostratus	卷层云
CMA	China Meteorological Administration	中国气象局
D	depression	低压
D	deviation	偏差,偏向
D	difference	差别,差
D	diffusion	扩散
D	dimensional	维的,因次的,尺寸的 (0-D, 1D, 2D 即无因次,1 维,2 维)
DALR	dry-adiabatic lapse rate	干绝热递减率
dB(db)	decibel	分贝
DBT(dbt)	dry-bulb temperature	干球温度
div.	divergence	辐散,散度
dm	dynamic metre	动力米
dn(dyn)	dyne	达因
DPT	dew-point temperature	露点温度
DSR	direct solar radiation	直接太阳辐射
1-DSS	one-dimensional steady-state model	一维定常模式
1DT	one-dimensional time-dependent model	一维时变模式
DUST	duststorm	尘暴
d. v.	dependent variable	因变数
DWR	daily weather report	每日天气报告
dy. (dym)	dynamic meter	动力米
DZ	drizzle	毛毛雨
E	east	东
E	energy	能量
e	exponential	指数,指数的,幂数的
EL	equilibrium level	平衡高度
Eq.	equator	赤道
exp.	exponent	指数
F	Fahrenheit	华氏
F	field	场
F	flux	流量,通量
F	force	力
F	frequency	频率
F	Froude number	弗罗德数

f.	foot	英尺
Fc	fracto-cumulus	碎积云
FL-BP	(BPF) filter-band pass	带通滤波器
FL-HP	(HPF) filter-high pass	高通滤波器
FL-LP	(LPF) filter-low pass	低通滤波器
Fn	fracto-nimbus	碎雨云
FNT	front	锋
FNTGNS	frontogenesis	锋生
FNTLYS	forntolysis	锋消
Fs	fracto-stratus	碎层云
ft.	feet	英尺
g	acceleration of gravity	重力加速度
g	gramme	克
g	gravity	重力
μg	microgram	微克
GARP	Global Atmospheric Research Program	全球大气研究计划
GFDL	Geophysical Fluid Dynamics Laboratory	地球物理流体动力学实验室
GMS	Geostationary Meteorological Satellite	地球静止气象卫星
GMT	Greenwich Mean Time	格林威治标准时
gpm	geopotential metre	位势米
grad.	gradient	梯度,坡度
h	height	高度
h	hour	小时
HP	high pressure	高压
HP	horizontal plane	水平面
HP	hydrostatic pressure	流体静压力
hPa	hectopascal	百帕
HPD	hourly precipitation data	每小时降水资料
hr.	hour	小时
HYO	half—yearly oscillation	半年振荡,半年一次涛动
IC	isobaric contours	等压线
in.	inch	英寸
ins	inches	英寸
I/O	input/output	输入/输出
IR	infrared absorption	红外线吸收
IR	infrared	红外的
IR	infrared radiation	线外辐射

ITCZ	intertropical convergence zone	热带辐合带
J(j)	joule	焦耳
Jan	January	一月
JAS	Journal of the Atmospheric Sciences	美《大气科学杂志》
K	Kelvin absolute scale	开氏绝对温标
k	Kilo-	千
K	knot	海里/小时=1.8532千米/小时
kg	kilogram	千克
kHz	kilohertz	千赫
km	kilometer	千米
km/h	kilometers per hour	千米/小时
km.p.h.	kilometers per hour	千米/小时
kn.	(kt) knot	海里/小时
kPa	kilopascal	千帕
kph	kilometers per hour	千米/小时
kW	kilowatt	千瓦
Lab. (lab)	laboratory	实验室
LADAR	laser detection and ranging (laser radar)	激光雷达
laf	laminar air flow	片流,片状气流
LAM	limited-area model	有限区域模式
Lat. ht.	latent heat	潜热
Lb. (lb.)	libra (pound)	磅
LCL	lifted (lifting) condensation level	抬升凝结高度
TLCM	seven-layer coarse mesh model	7层粗网格模式
LFC	level of free convection	自由对流高度
LFM	Limited-area fine mesh model	有限区域细网格模式
LFO	low frequency oscillation	低频振荡
12L GLO	Twelve-layer Global Model	12层全球模式
L. H. (l.h.)	latent heat	潜热
LHR	latent heat release	潜热释放
left	hand rule	左手定则
LIA	Little Ice Age	小冰期
lim.	limit	极限、限度
LLJ	low level jet	低空急流
LLWS	low level wind shear	低空风切变
L. M. T	local mean time	地方平均时,当地时间
LMW	level of maximum wind	最大风速高度

ln	logarithm, natural	自然对数
LND	level of nondivergence	无辐散高度
LNGM	Limited-Area Nested Grid Model	有限区域套网格模式
LOCAT	low-altitude clear air turbulence	低空晴空湍流
log	logarithm	常用对数
L. P.	low pressure	低压
7LPE	Seven-Layer Primitive Equation Model	7层原始方程模式
LRF	long-rang forecast	长期预报
LSQ	line-squall	线飑
LST	local solar time	地方太阳时
L. S. T	local standard time	地方标准时
LT	lag-time	时间滞后
L. T.	local time	地方时
L. W.	long wave	长波
LWC	liquid water content	液态含水量
ly	langley	兰利=1 卡/厘米^2
l. y.	light year	光年=9.4605×10^{12} 千米
m.	mass	质量
m.	meter	米
m.	mile	英里
MARS	meteorological automatic reporting station	自动气象报告站
mb(mB.)	millibar	毫巴
MCC	mesoscale convective complex	中尺度对流复合体
MCC	mesoscale cellular convection	中尺度环流对流
MCS	mesoscale convective system	中尺度对流系统
met.	meteorological(meteo, meteor, meteorol)	气象学的
mi.	mile	英里
min.	minimum	极小,最小值
mmHg	millimetre of mercury	毫米汞柱(高)
Mnsn	monsoon(MS)	季风
MONEX	Monsoon Experiment	季风试验(GARP)
M. W. R.	Monthly Weather Review	《每月天气评论》
MOS	model output statistics	模式输出统计预报
m. p. h	mile per hour	英里/小时
MPT	Mid-Pacific Trough	太平洋中部槽
MR	multiple regression	多重回归
MRA	maximum rain area	最大降水区

MRF	medium-range forecast	中期预报
m/s	meters per second	米/秒
MSM	Meso-scale Model	中尺度模式
N(Nb)	nimbus	雨云
N	north pole	北极
NASA	National Aeronautic and Space Administration	[美]国家航空航天局
NBL	nocturnal boundary layer	夜间边界层
NCAR	National Center for Atmospheric Research	[美]国家大气研究中心
NOAA	National Oceanic and Atmospheric Administration	[美]国家海洋大气局,〈诺阿〉气象卫星
NSSFC	National Severe Storm Forecast Center	[美]国家强风暴预报中心
NWP	numerical weather prediction	数值天气预报
NWS	National Weather Service	[美]国家气象局
OCLN	occlusion	锢囚
Oxy	oxygen	氧
OZ	ozone	臭氧
p	pressure	气压
p	precipitation	降水
p	probability	概率
PAM	portable automated meso-network	移动式中尺度观测网
PBL	planetary boundary layer	行星边界层
PDE	partial differential equation	偏微分方程
PE	potential energy	位能
PE	primitive equation	原始方程
phy[s].	physics	物理学
PMP	probable maximum precipitation	可能最大降水量
POP	probability of precipitation	降水概率[预报]
PP	perfect prognosis(PPM)	完全预报
PPI	plane position indicator	平面位置显示器
PPN	precipitation	降水
PPR	percentage probability of rain	降雨量百分概率
pres.	pressure	气压,压强
prob.	probability	概率
PROFS	Program for Regional Observing and Forecasting System	[美]区域观测和预报系统计划
prog	prognostic	预报的
PROMIS	program for an Operational Meteorological Information System	[瑞典]气象业务情报系统

PVA	positive vorticity advection	正涡度平流
PW	precipitable water	可降水分
p—wave	pressure wave	气压波
r	ratio	比率
r	radius	半径
rad	radian	弧度
REEP	regression estimation of event probabilities	事件概率的回归估计
R. H	relative humidity	相对湿度
RHI	range-height indication	距离—高度显示
RHI	range-height indicator	距离—高度显示器
RTE	radiative transfer equation	辐射传输方程
s	second	秒
s	speed	速率
sec.	second	秒
Sc	strato-cumulus	层积云
SESAME	Severe Environmental Storms and Mesoscale Experiment	强[环境]风暴和中尺度试验
SESAME	structure and energetics of the stratosphere and mesosphere	平流层与中间层大气结构和能量学
Sh.	shower	阵雨
SI	stability index	稳定度指数
SI	System International	[法]国际单位制
s. l.	sea level	海平面
SLBC	sea and land-breeze circulation	海陆风环流
SO	southern oscillation	南方涛动
SOR	successive over-relaxation	逐次超松弛法
SP	static pressure	静压力
SR	severe right moving storm	右移强风暴
SST	sea surface temperature	海面温度
SSTA	sea-surface temperature anomaly	海面温度异常
STCZ	sub-tropical convergence zone	副热带辐合区
STH	subtropical High	副热带高压
STJ	subtropical jet stream	副热带急流
STM	storm	风暴,暴风
STORM	storm-scale Operational and Research Meteorology	风暴尺度业务和研究气象学计划
STR	sub-tropical ridge	副热带高压脊
strato.	stratosphere	平流层

STRATWAR	Stratospheric warming	平流层增温
STS	severe tropical storm	强热带风暴
S. W.	short wave	短波
TC	tropical cyclone	热带气旋
TD	tropical depression	热带低压
temp.	temperature	温度
tropo	tropospheric	对流层的
TW	typhoon warning	台风警报
TWC	time weighted average	时间加权平均
twds	trade wind	信风
UCAR	University Corporation for Atmospheric Research	[美]大学大气研究协会
UHR	ultra-high resolution	超高分辨率
UN	United Nations	联合国
U. S.	United States	美国
vis.	visibility	能见度
VOR	vorticity	涡度
WBPT	wet-pulb potential temperature	湿球位温
wea.	weather	天气,气象
wea. obser.	weather observer	气象观测员
wfp	warm front passage	暖锋过境
WMF	warm front	暖锋
WMO	World Meteorological Organization	世界气象组织
WS	weather station	气象站
WS	wind speed	风速
wtd av.	weighted average	加权平均
WVF	water vapor flux	水汽通量
WVT	water vapor transmission	水汽输送
WWW	world weather watch	世界天气监视网
yr(yr.)	year	年
Z	Greenwich Mean Time	格林尼治平时

10.3 缩略语的读音规则

缩略语的读法有两种:

(1)按各字母组成的音节读。如:NCAR 读作"恩卡",UCAR 读作"由卡",MOS 读作"模斯"等;

(2)按字母名称读。如 WMO,ITCZ,MCC 等。

主要参考文献

陈忠美,等,1996.气象科技英语听说教程[M].北京:气象出版社.
郭坤,1984.科技英语写作入门[M].南京:江苏科学技术出版社.
李学平,1985.科技英语阅读新方法[M].广州:华南工学院出版社.
聂继武,等,1986.科技英语翻译教程[M].北京:机械工业出版社.
盛芝义,等,1992.科技英语900句[M].北京:气象出版社.
孙娴柔,1979.谈谈写作英语科技论文[M].北京:科学出版社.
谭丁,方苇,1980.科技英语读本[M].北京:农业出版社.
谭丁,盛芝义,1987.大气科学英语读物选[M].北京:气象出版社.
张义斌,等,1985.科技英语快速阅读技巧[M].武汉:华中工学院出版社.
赵志毅,1985.科技英语三百问[M].西安:陕西人民出版社.
Anthes R A, et al,1975. The Atmosphere[M]. printed in USA.
Houze Jr R A, 2004. Mesoscale Convective Systems[J]. Reviews of Geophysics, 42,2004RG,000150.
Houze Jr R A, 2014. Cloud Dynamics[M]. Second edition. International Geophysics Series, Volume 104, Academic Press.
Wallace J M, Hobbs P V, 2006. Atmospheric Science, An Introductory Survey[M]. Second Edition. Academic Press.
Yang J, Zhang Z Q, Wei C Y, et al, 2017. Introducing the New Generation of Chinese Geostationary Weather Satellites, Fengyun-4[J]. BAMS:1637-1658.